Transitioning to Calculus

Lawrence M. Leemis
Department of Mathematics
The College of William & Mary
Williamsburg, Virginia

Library of Congress Cataloging-in-Publication Data

Leemis, Lawrence M.
    Transitioning to Calculus / Lawrence M. Leemis.
    ISBN 978-0-9829174-9-7
    1. Mathematics–textbooks
    QA 303.2.L44 2012

© 2012 by Lawrence M. Leemis

All rights reserved. No part of this book may be reproduced in any form or by any means, without permission in writing from the publisher.

The author and publisher of this book have used their best efforts in preparing this book. These efforts include the checking of the mathematics for correctness. The author and publisher make no warranty of any kind, expressed or implied, with regard to the mathematics contained in this book. The author and publisher shall not be liable in any event for incidental or consequential damages in connection with, or arising out of, the furnishing, performance, or use of the mathematics contained herein.

Printed in the United States of America

10 9 8 7 6 5 4 3 2 1

ISBN 978-0-9829174-9-7

For Jill, Lindsey, Mark, and Logan

Contents

1 **Arithmetic**   1

2 **Algebra**   27

3 **Geometry**   63

4 **Analytic Geometry**   81

5 **Trigonometry**   133

6 **Complex Numbers**   173

Preface

If you were to ask ten calculus instructors what they expect their students to know prior to entering their calculus classes, you would get ten different answers. This book is designed to contain the average of those ten answers. Rather than trying to be a complete pre-calculus text, this book reviews only the essential concepts and formulas, with an occasional brief illustration.

The study of calculus involves a variety of new techniques for solving problems, but it rests on a wide array of skills that have been learned in the elementary and secondary grades. Students often have gaps in these skills due to switching of curriculum or changing schools. The purpose of this book is to provide a comprehensive list of the skills that are required of students entering their first class in calculus.

The formulas are given in outline form and are divided into the following six chapters: arithmetic, algebra, geometry, analytic geometry, trigonometry, and complex numbers. Make sure not to skip the arithmetic chapter. Although it will seem rather elementary at first, there are several concepts introduced that will be new to most readers. Index entries with page numbers set in italics indicate a primary source.

Exercises have been included at the end of each chapter. They are ordered from easier exercises to more difficult exercises. Each exercise appears on an odd-numbered page. The associated solution is given on the following even-numbered page. The reader is encouraged to solve the problem on the space provided in the book when an exercise is encountered, then check the solution given on the following page. The vast majority of the exercises cover necessary skills for a calculus course; a few of the exercises cover related ACT and SAT exam topics.

If this book is to be read during the summer prior to taking calculus during the fall, one approach would be to divide the number of days remaining until the first calculus class into the number of pages in the book and attempt to consistently cover that number of pages every day. There are 180 pages in the text by design. This means that the text can be read in one, two, or three months by reading a round number of pages every day.

My thanks goes to Dana Johnson, who suggested a book with this content, and to Nancy Corliss, John Drew, Ted Findler, Linda Hite, Bob Johnson, and Dave Lutzer, for their helpful comments on the text. My thanks also goes to Lindsey Leemis for the handsome cover design.

Best wishes to you in your first calculus course.

Williamsburg, Virginia   Larry Leemis
                                                                                                     July 2011

Chapter 1

Arithmetic

Arithmetic introduces the elementary mathematical operators (addition, subtraction, multiplication, and division), along with more advanced notions such as exponents, decimal notation, and percentages. Operations such as the absolute value and factorial are also defined here.

- Whole numbers

    - The *whole numbers* are created using the *digits* 0, 1, 2, 3, 4, 5, 6, 7, 8, 9
    - *Place value* is used to represent whole numbers exceeding nine. Reading from right to left, the meaning of the digits are ones, tens, hundreds, thousands, etc., for example,

    $$9876 = 9 \cdot 1000 + 8 \cdot 100 + 7 \cdot 10 + 6 \cdot 1$$

    which is read as "nine thousand, eight hundred, seventy-six."

    - For readability, commas are used to place digits into groups of three proceeding from right to left for whole numbers containing four or more digits, for example, the speed of light is approximately

    $$299,792,458$$

    meters per second

- Number lines

    - A *number line* typically places *negative numbers* to the left of 0 and *positive numbers* to the right of 0

    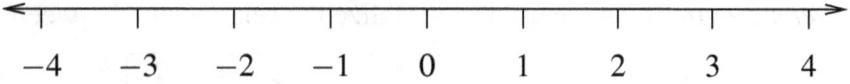

    - The point on a number line at 0 is known as the *origin*
    - The $+$ sign is generally dropped for brevity, for example, $+4$ is usually written as simply 4

- Two different numbers that are equal distance from the origin (for example, -3 and 3) are known as *opposites*
- The $<$ symbol is read "is less than," for example, $1 < 4$ means 1 lies to the left of 4 on the number line
- The $>$ symbol is read "is greater than," for example, $-2 > -4$ means that -2 lies to the right of -4 on the number line

- Common mathematical symbols
    - The $=$ symbol is read "equals" or "is equal to." Two quantities on either side of an equals sign have the same value.
    - The $\neq$ symbol is read "is not equal to," for example, $3 \neq 7$.
    - The $\pm$ symbol is read "plus or minus," for example, $x = \pm 3$ means that x is equal to 3 or x is equal to -3

- Sets
    - A *set* is a collection of objects
    - The objects that comprise a set are often known as *elements* of the set
    - Uppercase letters are typically used to denote sets
    - Sets are typically defined by placing their elements in curly braces, for example,
    $$A = \{\text{Cubs, Sox, Mets}\} \qquad B = \{1, 3, 5, 7\} \qquad C = \{\pm 1, \pm 3, \pm 5, \ldots\}$$
    where the ellipsis $(\ldots)$ indicates repetition (etc. or "and so on")
    - The number of elements in a set is known as its *cardinality*, for example, $N(A) = 3$, $N(B) = 4$, and $N(C)$ is "countably infinite"
    - If every element of a set B is also an element of a set C, as is true for the sets B and C defined above, then B is called a *subset* of C. This is denoted by: $B \subset C$.
    - The symbol $\in$ is read as "is an element of," for example, $3 \in C$
    - The ordering of elements in a set is not significant, for example, $\{2, 4, 6\} = \{6, 2, 4\}$
    - Two sets that have no elements in common are called *disjoint* or *mutually exclusive*
    - The *empty set* or *null set* $\emptyset$ is a set that contains no elements
    - *Venn diagrams* can be used to visualize the relationship between sets

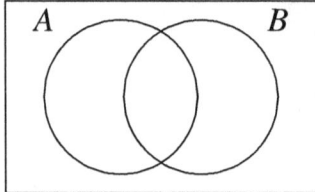

Chapter 1. Arithmetic

- Operations on sets
  * Union: $A \cup B$ is read "A union B" and is the set of all elements in either A or B

  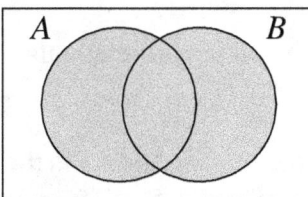

  * Intersection: $A \cap B$ is read "A intersect B" and is the set of all elements in both A and B

  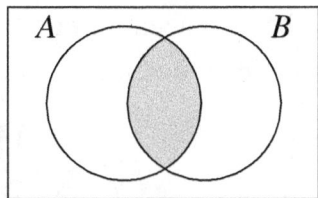

  * Complement: A' is read "A complement" and is the set of all elements *not* in A

  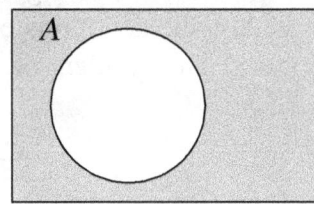

- The integers and some special subsets of the integers

$$\begin{aligned} \text{Integers:} \quad & Z = \{\ldots, -2, -1, 0, 1, 2, \ldots\} \\ \text{Natural numbers (positive integers):} \quad & N = Z^+ = \{1, 2, \ldots\} \\ \text{Whole numbers (nonnegative integers):} \quad & \{0, 1, 2, \ldots\} \\ \text{Negative integers:} \quad & Z^- = \{-1, -2, \ldots\} \\ \text{Even integers:} \quad & \{0, \pm 2, \pm 4, \ldots\} \\ \text{Odd integers:} \quad & \{\pm 1, \pm 3, \pm 5, \ldots\} \end{aligned}$$

- Basic arithmetic operations
  - Addition: "Two plus three equals five" is written as

  $$2 + 3 = 5$$

  where 2 and 3 are the *addends* and 5 is the *sum*. The number line interpretation of this addition problem is "beginning at 2 and moving 3 units to the right results in 5." Addition can be commuted, for example, $2 + 3 = 5$ and $3 + 2 = 5$.

  - Subtraction: "Five minus three equals two" is written as

  $$5 - 3 = 2$$

where 5 is the *minuend*, 3 is the *subtrahend*, and 2 is the *difference*. The number line interpretation of this subtraction problem is "beginning at 5 and moving 3 units to the left results in 2."

- Multiplication: "Five times three equals fifteen" is written as

$$5 \cdot 3 = 15 \qquad \text{or} \qquad 5 \times 3 = 15 \qquad \text{or} \qquad (5)(3) = 15$$

where 5 and 3 are the *factors* and 15 is the *product*. The interpretation of multiplication is repeated addition ($5+5+5 = 15$ or $3+3+3+3+3 = 15$). Multiplication can be commuted, for example, $5 \cdot 3 = 15$ and $3 \cdot 5 = 15$. The product of two negative numbers is a positive number, for example, $(-5)(-3) = 15$. The product of a positive number and a negative number is a negative number, for example, $(5)(-3) = (-5)(3) = -15$.

- Division: "Fifteen divided by three equals five" is written as

$$15 \div 3 = 5 \qquad \text{or} \qquad 15/3 = 5 \qquad \text{or} \qquad \frac{15}{3} = 5$$

where 15 is the *numerator* and 3 is the *denominator* and 5 is the *quotient* (or *ratio*). The interpretation of division is dividing the number in the numerator into the number of equal parts indicated by the denominator, then noting the size of the parts. Division by zero is not permitted. The quotient of two negative numbers is a positive number, for example,

$$\frac{-15}{-3} = 5$$

The quotient of a positive number and a negative number is a negative number, for example,

$$\frac{-15}{3} = \frac{15}{-3} = -5$$

The *remainder* is the amount remaining after the division of two integers, for example, when 17 is divided by 3, the remainder is 2.

- Multiples

    - A *multiple* is the result of multiplying a number by an integer
    - The *multiples* of 7 are 7, 14, 21, 28, 35, ...
    - The *least common multiple* (LCM) of two or more natural numbers is the smallest common multiple, for example, LCM(4, 6) = 12

- Factors

    - A *factor* is a natural number that can be evenly divided (divided with a remainder of zero) into another natural number. The *factors* of 24, for example, are 1, 2, 3, 4, 6, 8, 12, and 24
    - The *greatest common factor* (GCF) is the largest factor common to two or more numbers, for example, GCF(24, 36) = 12

Chapter 1. Arithmetic

- Prime numbers

  - *Prime numbers* have exactly two *distinct* factors, for example, 17 has factors 1 and 17
  - The first ten primes are 2, 3, 5, 7, 11, 13, 17, 19, 23, 29
  - There is no largest prime number
  - The *prime factorization* of an integer writes it as a product of primes, for example,

  $$120 = 2 \cdot 2 \cdot 2 \cdot 3 \cdot 5$$

  - *Composite numbers* are positive integers greater than 1 that are not prime
  - The first ten composite numbers are 4, 6, 8, 9, 10, 12, 14, 15, 16, 18

- Factorial: for a whole number n, the product of the consecutive integers from 1 to n is known as "n factorial" and is denoted by

$$n! = n \cdot (n-1) \cdot (n-2) \cdot \ldots \cdot 3 \cdot 2 \cdot 1$$

for example, $4! = 4 \cdot 3 \cdot 2 \cdot 1 = 24$. By definition, $0! = 1$.

- Modulo: the *modulo operator* returns the remainder when the second integer is divided into the first integer, for example,

$$57 \bmod 10 = 7$$
$$48 \bmod 8 = 0$$
$$225 \bmod 2 = 1$$

- Fractions

  - The fraction a/b is defined for any a and any $b \neq 0$
  - The fraction a/b is sometimes referred to as the *ratio* of a to b
  - If $a > b$ then the fraction a/b is an *improper fraction*, for example, 17/3
  - A *mixed number* is the sum of an integer and a fraction, for example, $2\frac{1}{5} = 2 + \frac{1}{5}$
  - To convert an improper fraction to a mixed number, divide the numerator by the denominator, which gives the whole number; the remainder in this division is the numerator of the fractional portion of the mixed number, and the denominator is the same as in the original improper fraction, for example,

  $$\frac{20}{3} = \frac{18}{3} + \frac{2}{3} = 6\frac{2}{3}$$

  - To convert a mixed number to an improper fraction, multiply the denominator by the whole number, then add the numerator, which gives the numerator of the improper fraction. The denominator stays the same, for example,

  $$6\frac{2}{3} = \frac{3 \cdot 6 + 2}{3} = \frac{20}{3}$$

- Basic arithmetic operations
  * Addition (identical denominators)
  $$\frac{a}{b} + \frac{c}{b} = \frac{a+c}{b}$$
  * Addition (different denominators)
  $$\frac{a}{b} + \frac{c}{d} = \frac{ad+bc}{bd}$$
  * Subtraction (identical denominators)
  $$\frac{a}{b} - \frac{c}{b} = \frac{a-c}{b}$$
  * Subtraction (different denominators)
  $$\frac{a}{b} - \frac{c}{d} = \frac{ad-bc}{bd}$$
  * Multiplication
  $$\frac{a}{b} \cdot \frac{c}{d} = \frac{ac}{bd}$$
  * Division ("invert and multiply")
  $$\frac{a}{b} \div \frac{c}{d} = \frac{a}{b} \cdot \frac{d}{c} = \frac{ad}{bc}$$
  or, equivalently
  $$\frac{\frac{a}{b}}{\frac{c}{d}} = \frac{a}{b} \cdot \frac{d}{c} = \frac{ad}{bc}$$

- A fraction can be expressed in *lowest terms* by writing the *prime factorization* of the numerator and denominator and canceling equal factors, for example,
$$\frac{24}{36} = \frac{\cancel{2} \cdot \cancel{2} \cdot 2 \cdot \cancel{3}}{\cancel{2} \cdot \cancel{2} \cdot 3 \cdot \cancel{3}} = \frac{2}{3}$$

- Fractions can also be reduced to *lowest terms* by dividing the numerator and the denominator by the greatest common factor of the numerator and denominator, for example,
$$\frac{24}{36} = \frac{24 \div 12}{36 \div 12} = \frac{2}{3}$$

- The *least common denominator* (LCD) is the smallest positive integer divisible by each of the denominators of a group of fractions, for example, the least common denominator of the fractions
$$\frac{1}{3}, \frac{5}{6}, \frac{2}{9}$$

Chapter 1. Arithmetic

is 18. This is useful for adding and subtracting fractions, for example,

$$\frac{1}{3} + \frac{5}{6} - \frac{2}{9} = \frac{6}{18} + \frac{15}{18} - \frac{4}{18} = \frac{6+15-4}{18} = \frac{17}{18}$$

The LCD of a group of fractions is the LCM of their denominators.

- Two numbers are *reciprocals* if their product is 1, for example, 5/7 and 7/5 are reciprocals, -3 and $-1/3$ are reciprocals

- Rational numbers, irrational numbers, and real numbers
  - A *rational number* is the ratio of two integers (the denominator can't be zero), for example, 7/4, 17, $-2/3$, 0
  - The set of all rational numbers is denoted by Q, as in Quotient
  - An *irrational number* is a real number that is not rational, for example, $\sqrt{17}$
  - Two important irrational numbers are

    $$\pi = 3.141592653589793\ldots \quad \text{and} \quad e = 2.718281828459045\ldots$$

  - The *real numbers* are comprised of the rational and irrational numbers
  - A *real number* can be expressed as an infinite decimal
  - The set of all real numbers is denoted by $\mathbb{R}$

- Decimals
  - A number expressed as a *decimal* often results from performing a division
  - Whole numbers are placed to the left of the *decimal point* (which is formatted as a period); fractional values are placed to the right of the decimal point
  - The digits to the left of the decimal point (reading right to left) are ones, tens, hundreds, thousands, ...
  - The digits to the right of the decimal point (reading left to right) are tenths, hundredths, thousandths, ...
  - Examples of decimals:

    $$21\frac{3}{10} = 21.3 \qquad 757\frac{38}{100} = 757.38 \qquad \frac{7}{100} = 0.07$$

    Including the leading 0 on 0.07 is optional, that is, 0.07 and .07 are acceptable ways to write the decimal

  - A bar is used to denote repeating digits, for example,

    $$1/22 = 0.045454545\ldots = 0.0\overline{45}$$

- A rational number a/b can be expressed as a *terminating decimal* if, in the long division process, one eventually gets a remainder of 0, for example,

$$1/5 = 0.2 \qquad 11/8 = 1.375$$

- A rational number a/b can be expressed as a *repeating decimal* if b divides a so that the decimal has a repeating pattern of integer digits, for example,

$$1/3 = 0.3333\ldots \qquad 2/7 = 0.\overline{285714}$$

- Rational numbers can be written as terminating decimals or repeating decimals
- Irrational numbers have a decimal form that is non-terminating and non-repeating, for example,

$$\pi = 3.141592653589793\ldots \qquad \sqrt{2} = 1.414213562373095\ldots$$

- A decimal can be *truncated* by eliminating certain right-hand digits, for example, 365.242199 truncated to four decimal places beyond the decimal point is 365.2421
- A decimal can be *rounded* by eliminating certain right-hand digits after inspecting the left-most digit eliminated, for example, 365.242199 rounded to four decimal places beyond the decimal point is 365.2422

- Rounding
  - Rounding a numerical value is the process of finding an approximately equal value that has a simpler expression to a prescribed level of accuracy
    * 337 rounded to the nearest 10 is "rounded up" to 340
    * 337 rounded to the nearest 100 is "rounded down" to 300
    * 8/3 rounded to the nearest integer is 3
    * 8/3 rounded to the nearest hundredth is 2.67
    * $12.3457 rounded to the nearest penny is $12.35
    * $12.35 rounded to the nearest dollar is $12
    * π rounded to the nearest hundred thousandth is 3.14159
    * π rounded to the nearest ten thousandth is 3.1416
  - When the numerical value being rounded falls exactly halfway between two rounded values, one common practice is to round up
    * 42.5 rounded to the nearest integer is 43
    * -42.5 rounded to the nearest integer is -42
  - Rounding is useful for quickly determining an approximate solution to an arithmetic problem, for example, the sum of 4017 and 2976 is *approximately* $4000 + 3000 = 7000$

Chapter 1. Arithmetic

- Absolute value
    - The *absolute value* of a real number a is the distance between a and the origin on a number line. Equivalently, the absolute value is the numerical part of a number, ignoring its sign.
    - For any real number a, the absolute value of a is
    $$|a| = \begin{cases} a & \text{if } a \geq 0 \\ -a & \text{if } a < 0 \end{cases}$$
    - The absolute value of $a - b$, denoted $|a - b|$, is the distance between a and b on a number line
    - Absolute value properties
        * $|a| \geq 0$
        * $|-a| = |a|$
        * $|a - b| = |b - a|$
        * $|ab| = |a| \cdot |b|$
        * $\left|\dfrac{a}{b}\right| = \dfrac{|a|}{|b|}$

- Floor and ceiling functions
    - The floor and ceiling functions convert real numbers to nearby integers
    - The *floor function*, also known as the *greatest integer function*, returns the largest integer that is less than or equal to its argument, for example,
    $$\lfloor 3.7 \rfloor = 3 \qquad \left\lfloor \frac{17}{2} \right\rfloor = 8 \qquad \lfloor 17 \rfloor = 17 \qquad \lfloor -5.2 \rfloor = -6$$
    - The *ceiling function*, also known as the *least integer function*, returns the smallest integer that is greater than or equal to its argument, for example,
    $$\lceil 3.7 \rceil = 4 \qquad \left\lceil \frac{17}{2} \right\rceil = 9 \qquad \lceil 17 \rceil = 17 \qquad \lceil -5.2 \rceil = -5$$

- Percentages
    - A percentage is a part of a whole expressed in hundredths, for example, the fraction $17/100$, the decimal 0.17, and the percentage 17% are all equal
    - To convert a fraction to a percentage, multiply by 100
    - To convert a percentage to a fraction, divide by 100
    - Use fractions or decimals when finding percentages, for example, to find 30% of 200, compute $(0.3)(200) = 60$

- The *percentage change* in a quantity is the difference between the final and original amounts, divided by the original amount. The change from $100 to $120, for example, is

$$\frac{\text{final} - \text{original}}{\text{original}} = \frac{120 - 100}{100} = \frac{20}{100} = 0.2$$

which corresponds to a 20% increase. The change from $100 to $80 is

$$\frac{\text{final} - \text{original}}{\text{original}} = \frac{80 - 100}{100} = -\frac{20}{100} = -0.2$$

which corresponds to a 20% decrease.

- Exponents

    - The expression x^n means multiply x by itself n times, provided that n is a natural number, for example, $5^3 = 5 \cdot 5 \cdot 5 = 125$
    - In the expression x^n, x is the *base*
    - In the expression x^n, n is the *exponent* or *power*
    - The n in the expression x^n is typeset as a *superscript*
    - Exponents of two and three are termed "squared" and "cubed"
        * The expression 5^2 is read "five squared"
        * The expression 7^3 is read "seven cubed"
        * The expression 10^4 is read "ten raised to the fourth power"
    - A leading negative sign is not included in the exponentiation unless parentheses are included, for example,

    $$-3^2 = -9 \qquad (-3)^2 = 9$$

    - Exponents can be used to write a prime factorization more compactly, for example

    $$120 = 2 \cdot 2 \cdot 2 \cdot 3 \cdot 5 = 2^3 \cdot 3 \cdot 5$$

    - The *perfect squares* 1, 4, 9, 16, 25, ... are the squares of the positive integers. These correspond to the areas of squares with integer side lengths.

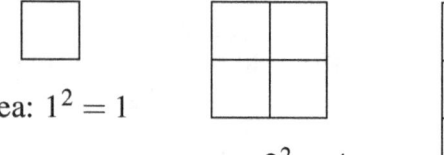

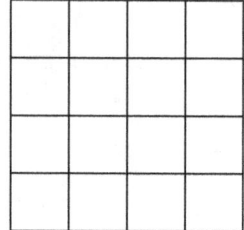

area: $1^2 = 1$

area: $2^2 = 4$

area: $3^2 = 9$

area: $4^2 = 16$

Chapter 1. Arithmetic

- The *perfect cubes* 1, 8, 27, 64, 125, ... are the cubes of the positive integers. These correspond to the volumes of cubes with integer side lengths.

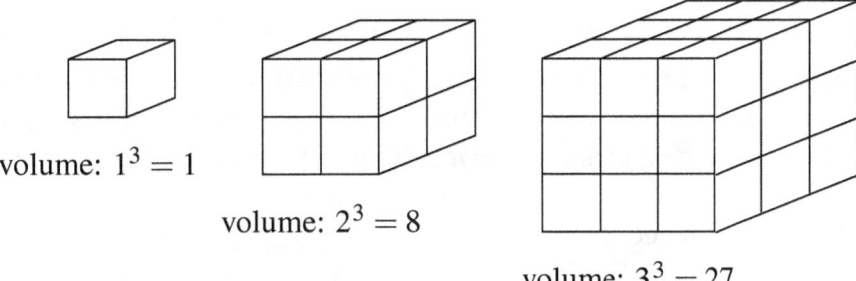

- Radicals
  - The square root of x, written as $\sqrt{x}$, denotes the number which, when multiplied by itself, equals x, for example, $\sqrt{16} = 4$
  - The cube root of x, written as $\sqrt[3]{x}$, denotes the number which, when multiplied by itself three times, equals x, for example, $\sqrt[3]{-8} = -2$
  - A radical implies a single *principal root*, for example, $\sqrt{9} = 3$, not ± 3

- *Measurement systems:* make sure units cancel properly when converting from one unit of measure to another, for example, the number of seconds in a non-leap year is

$$365 \text{ days} \times \frac{24 \text{ hours}}{\text{day}} \times \frac{60 \text{ minutes}}{\text{hour}} \times \frac{60 \text{ seconds}}{\text{minute}} = 31{,}536{,}000 \text{ seconds}$$

The factors that are fractions which are used to convert units are known as *unit multipliers*.

- Order of operations ("Please Excuse My Dear Aunt Sally" or PEMDAS): The priority order of operations when evaluating a complicated expression is
  - Parentheses (or brackets [·] or braces {·})
  - Exponents
  - Multiplication, Division
  - Addition, Subtraction

Work from left to right for operations with the same priority, for example,

$$\begin{aligned}
(2+3) \cdot 4 + 6^2/9 - 5 &= 5 \cdot 4 + 6^2/9 - 5 && \text{parentheses} \\
&= 5 \cdot 4 + 36/9 - 5 && \text{exponents} \\
&= 20 + 36/9 - 5 && \text{multiplication} \\
&= 20 + 4 - 5 && \text{division} \\
&= 24 - 5 && \text{addition} \\
&= 19 && \text{subtraction}
\end{aligned}$$

- Scientific notation

  - Very large and very small numbers are efficiently written using *scientific notation*, which is helpful for avoiding writing all of the leading or trailing zeros
  - Normalized form: $\pm a \times 10^n$, where $1 \leq a < 10$ and n is an integer (the 10 and exponent are typically omitted when $n = 0$); $n < 0$ for very small numbers (negative exponents are defined in the next chapter) and $n > 0$ for very large numbers, for example, 6.02×10^{23}

- Summation notation and formulas

  - The summation symbol $\sum$ includes an index (which is x in the example below); the lower limit of the index is given below the summation symbol; the upper limit is given above the summation symbol, for example,

  $$\sum_{x=3}^{5} \sqrt{2x} = \sqrt{6} + \sqrt{8} + \sqrt{10}$$

  - One can show that the sum of the first n positive integers is

  $$\sum_{x=1}^{n} x = 1 + 2 + \cdots + n = \frac{n(n+1)}{2}$$

  - One can show that the sum of the squares of the first n positive integers is

  $$\sum_{x=1}^{n} x^2 = 1^2 + 2^2 + \cdots + n^2 = \frac{n(n+1)(2n+1)}{6}$$

- Arithmetic sequences and series

  - A *sequence* is a list of numbers (referred to as *terms*) in a specific order
  - The terms in a sequence are traditionally separated by commas, for example,

  $$1, 4, 9, 16, 25, 36, \ldots$$

  - An *arithmetic sequence* is a list of numbers in a specific order with the difference between any two consecutive terms always the same, for example,

  $$1, 3, 5, 7, 9, 11, \ldots \text{(common difference between consecutive terms is 2)}$$

  or

  $$8, 5, 2, -1, -4, -7, \ldots \text{(common difference between consecutive terms is } -3)$$

  - The terms in an arithmetic sequence are denoted by $a_1, a_2, a_3, \ldots$

Chapter 1. Arithmetic

- The integers that index the *a* values are known as *subscripts*
- The *common difference* between consecutive terms in an arithmetic sequence is denoted by *d*
- A *recursive formula* for finding the *n*th term in an arithmetic sequence is

$$a_n = a_{n-1} + d$$

- A *general formula* for finding the *n*th term in an arithmetic sequence is

$$a_n = a_1 + (n-1)d$$

- An *arithmetic series* is a running sum of the terms in an arithmetic sequence, for example, the sum of the first four terms in the arithmetic sequence $1, 3, 5, 7, 9, 11, \ldots$ is

$$1 + 3 + 5 + 7 = 16$$

- The sum of the first *n* terms in an arithmetic sequence is

$$\begin{aligned} \sum_{i=1}^{n} a_i &= a_1 + a_2 + a_3 + \cdots + a_n \\ &= a_1 + (a_1 + d) + (a_1 + 2d) + \cdots + (a_1 + (n-1)d) \\ &= \frac{n(a_1 + a_n)}{2} \end{aligned}$$

- Geometric sequences and series

  - A *geometric sequence* is a list of numbers in a specific order with the ratio between any two consecutive terms always the same, for example,

    $3, 6, 12, 24, 48, 96, \ldots$ (common ratio between consecutive terms is 2)

    or

    $27, 9, 3, 1, \dfrac{1}{3}, \dfrac{1}{9}, \ldots$ (common ratio between consecutive terms is 1/3)

  - The terms in a geometric sequence are denoted by $a_1, a_2, a_3, \ldots$
  - The *common ratio* between consecutive terms in a geometric sequence is denoted by *r*
  - A *recursive formula* for finding the *n*th term in a geometric sequence is

    $$a_n = a_{n-1} r$$

  - A *general formula* for finding the *n*th term in a geometric sequence is

    $$a_n = a_1 r^{n-1}$$

- A *geometric series* is a running sum of the terms in a geometric sequence, for example, the sum of the first four terms in the geometric sequence 3, 6, 12, 24, 48, 96, ... is

$$3 + 6 + 12 + 24 = 45$$

- The sum of the first n terms in a geometric sequence with $r \neq 1$ is

$$\begin{aligned} \sum_{i=1}^{n} a_i &= a_1 + a_2 + a_3 + \cdots + a_n \\ &= a_1 + a_1 r + a_1 r^2 + \cdots + a_1 r^{n-1} \\ &= \frac{a_1(1-r^n)}{1-r} \end{aligned}$$

- When the common ratio r lies between -1 and 1, the *infinite geometric series* converges to

$$\sum_{i=1}^{\infty} a_i = \frac{a_1}{1-r} \qquad |r| < 1$$

- Counting techniques

    - *Multiplication Rule (Fundamental Theorem of Counting):* If there are n_1 ways to make decision 1, n_2 ways to make decision 2, ..., n_k ways to make decision k, then there are $n_1 \cdot n_2 \cdot \ldots \cdot n_k$ ways to make all of the decisions

    - *Permutations:* An arrangement of r objects taken from a set of n objects in a definite order is a *permutation*. The number of such permutations is

$$n \cdot (n-1) \cdot (n-2) \cdot \ldots \cdot (n-r+1) = \frac{n!}{(n-r)!}$$

    - *Combinations:* An arrangement of r objects taken from a set of n objects is a *combination*. The number of such combinations is

$$\binom{n}{r} = \frac{n!}{r!(n-r)!}$$

- Digits

    - The *Hindu–Arabic* numerals currently in use appear to have evolved from the *Brahmi* numerals

    - The first six numerals appear to be a collection of strokes consistent with the number that they represent: １ ⼆ ≡ ⼵ ち ʦ

    - The symbols 0, 1, 2, 3, 4, 5, 6, 7, 8, 9 are known as *digits*

Chapter 1. Arithmetic

- Number systems

    - The *decimal* number system, also known as the base 10 number system, uses digits 0, 1, 2, 3, 4, 5, 6, 7, 8, 9

        * The choice of 10 as a base corresponds with the number of digits (fingers and thumbs) on both hands
        * *Place value* and the use of a sign ($-$ for negative numbers) allows any number to be written in decimal using a sequence of digits, for example, denoting 365 in the decimal system as 365_{10},

        $$365_{10} = 3 \cdot 10^2 + 6 \cdot 10^1 + 5 \cdot 10^0 = 3 \cdot 100 + 6 \cdot 10 + 5 \cdot 1$$

        where the subscript 10 on 365 denotes the base

    - The *octal* number system, also known as the base 8 number system, uses digits 0, 1, 2, 3, 4, 5, 6, 7

        * The choice of 8 as a base corresponds to ignoring opposing thumbs and just using the eight fingers
        * Octal is a useful number system for digital computers
        * *Place value* and the use of a sign ($-$ for negative numbers) allows any number to be written in octal, for example 365_{10} can be written in octal as

        $$555_8 = 5 \cdot 8^2 + 5 \cdot 8^1 + 5 \cdot 8^0 = 5 \cdot 64 + 5 \cdot 8 + 5 \cdot 1$$

    - The *binary* number system, also known as the base 2 number system, uses digits 0 and 1

        * The choice of 2 as a base corresponds to counting arms or legs, rather than fingers and thumbs
        * Binary is a useful number system for digital computers
        * *Place value* and the use of a sign ($-$ for negative numbers) allows any number to be written in binary, for example $365_{10} = 555_8$ can be written in binary as

        $$\begin{aligned} 101101101_2 &= 1 \cdot 2^8 + 0 \cdot 2^7 + 1 \cdot 2^6 + 1 \cdot 2^5 + 0 \cdot 2^4 + 1 \cdot 2^3 + 1 \cdot 2^2 + 0 \cdot 2^1 + 1 \cdot 2^0 \\ &= 1 \cdot 256 + 0 \cdot 128 + 1 \cdot 64 + 1 \cdot 32 + 0 \cdot 16 + 1 \cdot 8 + 1 \cdot 4 + 0 \cdot 2 + 1 \cdot 1 \end{aligned}$$

        * The digits in binary are known as *bits* when stored on a computer
        * It is easy to convert between octal and binary because one octal digit corresponds to three consecutive binary digits

    - Roman numerals comprise another number system that was supplanted by the Hindu–Arabic numerals. The meaning of the symbols is given in the table below.

Roman numeral symbol	I	V	X	L	C	D	M
Value	1	5	10	50	100	500	1000

* There is no zero in the Roman numeral system
* The convention is to place the Roman numerals in descending order, for example, MMXII is $1000 + 1000 + 10 + 1 + 1 = 2012$
* When a lesser-value Roman numeral precedes a greater-value Roman numeral, the lesser value is subtracted from the greater value. For example, MMCMXLIV is

$$1000 + 1000 + (1000 - 100) + (50 - 10) + (5 - 1) = 2000 + 900 + 40 + 4 = 2944$$

* The first ten Roman numerals are I, II, III, IV, V, VI, VII, VIII, IX, X

Exercises

1.1 What is the prime factorization of 140?

1.2 Express $\frac{5}{6} + \frac{7}{9} - \frac{3}{2}$ as a single fraction.

1.3 Calculate $3(4-7) + |16/(6-8)|$.

1.4 Jill earns 10% in a stock mutual fund during a one-year period, then moves her funds to a second stock mutual fund where she earns 30% during the second year. What is Jill's overall rate of return for the two years? How does her overall rate of return compare to Logan's, if Logan earned 30% during the first year and 10% in the second year?

1.1 What is the prime factorization of 140?

The prime factorization of 140 is
$$140 = 2 \cdot 2 \cdot 5 \cdot 7.$$

1.2 Express $\dfrac{5}{6} + \dfrac{7}{9} - \dfrac{3}{2}$ as a single fraction.

The first step to adding these fractions is to find the least common denominator.

- The multiples of 6 are 6, 12, 18, 24, ...
- The multiples of 9 are 9, 18, 27, 36, ...
- The multiples of 2 are 2, 4, 6, 8, ...

The least common multiple of 6, 9, and 2 is LCM(6, 9, 2) = 18, which is the least common denominator. Converting each of these fractions so that they each have denominator 18 results in
$$\frac{5}{6} + \frac{7}{9} - \frac{3}{2} = \frac{15}{18} + \frac{14}{18} - \frac{27}{18} = \frac{15+14-27}{18} = \frac{2}{18} = \frac{1}{9}.$$

1.3 Calculate $3(4-7) + |16/(6-8)|$.

First perform the operations within parentheses, then perform the multiplication and division, then apply the absolute value operator, and finally perform the addition. The steps are shown below.
$$\begin{aligned} 3(4-7) + |16/(6-8)| &= (3)(-3) + |16/(-2)| \\ &= -9 + |-8| \\ &= -9 + 8 \\ &= -1. \end{aligned}$$

1.4 Jill earns 10% in a stock mutual fund during a one-year period, then moves her funds to a second stock mutual fund where she earns 30% during the second year. What is Jill's overall rate of return for the two years? How does her overall rate of return compare to Logan's, if Logan earned 30% during the first year and 10% in the second year?

Since the answer does not depend on how much money Jill invests, it is acceptable to assume that she starts with $100. Jill earns $(0.1)(\$100) = \$10$ during the first year, so she withdraws $110 from the first stock mutual fund. She earns $(0.3)(\$110) = \$33$ during the second year, bringing her total to $\$110 + \$33 = \$143$. So the overall rate of return for the two years is 43%. The common error here is to simply add 10% and 30%, but this ignores the effect of *compounding*, which is making a return during the second year on the return that she made during the first year. Using the same logic on Logan's investment, he earns $(0.3)(\$100) = \$30$ at the end of the first year, and earns $(0.1)(\$130) = \$13$ during the second year, bringing his total to $\$130 + \$13 = \$143$. Their rates of return are identical.

Chapter 1. Arithmetic

1.5 What are the three numbers that should be placed in the blanks below so that the difference between consecutive numbers in the sequence is the same?

$$19, \_\_, \_\_, \_\_, 75$$

1.6 Express $3.\overline{87}$ as a ratio of integers.

1.7 What is the last digit in 3^{999}?

1.5 What are the three numbers that should be placed in the blanks below so that the difference between consecutive numbers in the sequence is the same?

$$19, \underline{\quad}, \underline{\quad}, \underline{\quad}, 75$$

Since the difference between consecutive numbers in the sequence is the same, this is an arithmetic sequence. Using the arithmetic sequence notation, the first term is $a_1 = 19$. In order to find the common difference d, the fifth term of the arithmetic sequence can be expressed as

$$a_5 = a_1 + (5-1)d$$

or

$$75 = 19 + 4d$$

Solving this expression for d gives the common difference $d = 14$. The missing terms are 33, 47, and 61.

1.6 Express $3.\overline{87}$ as a ratio of integers.

Let x denote the fraction of interest, that is

$$x = 3.\overline{87}.$$

Since there are two repeating digits in $3.\overline{87}$, both sides of this equation should be multiplied by $10^2 = 100$:

$$100x = 387.\overline{87}.$$

When the first equation is subtracted from the second equation, the repeating portions of the numbers that lie to the right of the decimal point cancel, resulting in

$$100x - x = 387.\overline{87} - 3.\overline{87} = 384$$

or

$$99x = 384.$$

Finally, dividing by 99, the original repeating decimal can be expressed as

$$x = \frac{384}{99}.$$

1.7 What is the last digit in 3^{999}?

A calculator is not of use in solving this problem because of overflow. One approach is to write out the first few powers of 3:

$$3^0 = 1, \quad 3^1 = 3, \quad 3^2 = 9, \quad 3^3 = 27, \quad 3^4 = 81, \quad 3^5 = 243, \quad \ldots.$$

Examining the last digits, one can conclude that they fall in a cycle with a period of 4 (that is, 1, 3, 9, 7, 1, 3, 9, 7, 1, ...). This means that every exponent that is evenly divisible by 4 (that is, 0, 4, 8, ...) is associated with a power of 3 that has last digit 1. Thus 3^{1000} has last digit 1. Moving back one in the sequence, the last digit in 3^{999} is 7.

1.8 A restaurant offers three appetizers, four entrees, and two desserts. How many ways are there to order one appetizer, one entree, and one dessert?

1.9 If 140% of a number is 280, then what is 60% of that number?

1.10 How many of the first 1000 positive integers are multiples of neither 6 nor 9?

1.8 A restaurant offers three appetizers, four entrees, and two desserts. How many ways are there to order one appetizer, one entree, and one dessert?

Using the multiplication rule, there are $3 \cdot 4 \cdot 2 = 24$ different ways to order one appetizer, one entree, and one dessert.

1.9 If 140% of a number is 280, then what is 60% of that number?

Since 140% of the number is 280, the number must be

$$\frac{280}{1.4} = 200$$

because 140% is equivalent to 1.4. Finally, 60% of 200 is

$$(0.6)(200) = 120$$

because 60% is equivalent to 0.6.

1.10 How many of the first 1000 positive integers are multiples of neither 6 nor 9?

There are 166 multiples of 6, which are

$$6, 12, 18, \ldots, 996$$

because $166 \cdot 6 = 996$. Likewise, there are 111 multiples of 9, which are

$$9, 18, 27, \ldots, 999$$

because $111 \cdot 9 = 999$. An integer is a multiple of both 6 and 9 if it is a multiple of the least common multiple of 6 and 9, which is $\text{LCM}(6, 9) = 18$. The 55 integers between 1 and 1000 that are multiples of both 6 and 9 are

$$18, 36, 54, \ldots, 990$$

because $55 \cdot 18 = 990$. Letting the set A denote the multiples of 6 and the set B denote the multiples of 9, the Venn diagram below shows the counts of the various numbers of integers in the four regions partitioned by the sets A and B.

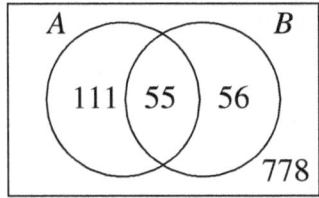

To answer the original question, there are 778 integers between 1 and 1000 that are multiples of neither 6 nor 9.

1.11 How many ways are there to select a committee of three people from a group of ten people?

1.12 How many of the first 5000 positive integers are neither perfect squares nor perfect cubes?

1.11 How many ways are there to select a committee of three people from a group of ten people?

Since the order that the people are selected for the committee is not important, combinations are used to conclude that there are

$$\binom{10}{3} = \frac{10!}{3!7!} = \frac{10 \cdot 9 \cdot 8}{3 \cdot 2 \cdot 1} = \frac{720}{6} = 120$$

different committees that can be selected.

1.12 How many of the first 5000 positive integers are neither perfect squares nor perfect cubes?

Let S be the set of all perfect squares in the first 5000 positive integers and let C be the set of all perfect cubes in the first 5000 positive integers. The perfect squares are

$$1, 4, 9, 16, \ldots.$$

The number of perfect squares in the first 5000 positive integers is

$$N(S) = \lfloor \sqrt{5000} \rfloor = 70.$$

The perfect cubes are

$$1, 8, 27, 64, \ldots.$$

The number of perfect cubes in the first 5000 positive integers is

$$N(C) = \lfloor \sqrt[3]{5000} \rfloor = 17.$$

The positive integers that are both perfect squares and cubes are

$$1, 64, 729, 4096, \ldots,$$

that is, the positive integers raised to the sixth power. The number of perfect cubes in the first 5000 positive integers is

$$N(S \cap C) = \lfloor \sqrt[6]{5000} \rfloor = 4.$$

Using a well-known result from set theory, the number of elements of $S \cup C$ is

$$N(S \cup C) = N(S) + N(C) - N(S \cap C) = 70 + 17 - 4 = 83.$$

The number of integers between 1 and 5000 (inclusive) that are neither perfect squares nor perfect cubes is

$$N(S' \cap C') = 5000 - N(S \cup C) = 5000 - 83 = 4917.$$

1.13 Find the number of distinct positive integer-valued solutions to the equation
$$a+b+c=9.$$

1.14 Find the number of trailing 0's at the end of 1000!.

1.13 Find the number of distinct positive integer-valued solutions to the equation
$$a+b+c=9.$$

One rather tedious solution to this problem is to enumerate all of the possible solutions, that is
$$a=1, b=1, c=7,$$
$$a=1, b=7, c=1,$$
etc. A more elegant and efficient solution to the problem is to frame the question in terms of nine balls ($\bullet, \bullet, \bullet, \bullet, \bullet, \bullet, \bullet, \bullet, \bullet$) and two dividers ($|, |$). The solution to the equation corresponding to the arrangement

$$\bullet\bullet\bullet\,|\,\bullet\bullet\bullet\bullet\bullet\,|\,\bullet$$

for example, is
$$a=3, b=5, c=1,$$
because there are three balls to the left of the first divider, five balls between the two dividers and one ball to the right of the second divider. Since there are eight gaps between the balls in which to place the two dividers and the dividers are indistinguishable, there are
$$\binom{8}{2} = \frac{8 \cdot 7}{2 \cdot 1} = \frac{56}{2} = 28$$
different ways to arrange the bars between the balls. Therefore, there are 28 different positive integer-valued solutions to the equation $a+b+c=9$.

1.14 Find the number of trailing 0's at the end of 1000!.

A calculator is not of use here due to overflow. Consider the prime factorization of 1000!. Every trailing zero corresponds to a five and two pair that appears in the prime factorization of 1000!. Since there are more twos than fives in the prime factorization, the problem reduces to finding the number of fives in the prime factorization of 1000!. There is

- one five in the factors 5, 10, 15, 20, 25, 30, ..., 1000,
- a second five in the factors 25, 50, 75, 100, 125, 150, ..., 1000,
- a third five in the factors 125, 250, 375, 500, 625, 750, 875, 1000,
- a fourth five in the factor 625.

Thus, the number of trailing 0's at the end of 1000! is
$$\left\lfloor \frac{1000}{5} \right\rfloor + \left\lfloor \frac{1000}{5^2} \right\rfloor + \left\lfloor \frac{1000}{5^3} \right\rfloor + \left\lfloor \frac{1000}{5^4} \right\rfloor,$$
which is the number of occurrences of the factor 5 in the prime factorization. This simplifies to
$$200+40+8+1=249$$
trailing zeros.

Chapter 2

Algebra

Elementary algebra introduces the notion of representing an unspecified number by a variable, which is typically denoted by a letter. The operations from arithmetic can be applied to these variables, oftentimes for the purpose of solving an equation. Word problems, which typically address problems of a practical nature, are typically formulated in terms of variables. Polynomials play a central role in algebra. Algebra begins with the rules of arithmetic in symbolic form.

- Fundamental properties of addition

    - Commutative property of addition: $a+b = b+a$
    - Associative property of addition: $(a+b)+c = a+(b+c)$
    - Distributive property of addition: $a(b+c) = ab+ac$
    - Additive identity property of addition: $a+0 = a$
    - Additive inverse property of addition: $a+(-a) = 0$

- Fundamental properties of multiplication

    - Commutative property of multiplication: $a \cdot b = b \cdot a$
    - Associative property of multiplication: $(a \cdot b)c = a(b \cdot c)$
    - Multiplicative identity property of multiplication: $a \cdot 1 = a$
    - Multiplicative inverse property of multiplication: $a \cdot \frac{1}{a} = 1, a \neq 0$

- Exponents

    - The expression x^a is "x raised to the a power," where x is the *base* and a is the *exponent* or *power*

– Rules for exponents

$$x^0 = 1 \quad \text{for } x \neq 0 \ (0^0 \text{ is not defined})$$
$$x^a x^b = x^{a+b} \quad \text{like base, add exponents}$$
$$\frac{x^a}{x^b} = x^{a-b}$$
$$(x^a)^b = x^{ab}$$
$$(xy)^a = x^a y^a$$
$$\left(\frac{x}{y}\right)^a = \frac{x^a}{y^a} \quad \text{for } y \neq 0$$
$$x^{-a} = \frac{1}{x^a} \quad \text{for } x \neq 0$$

- Radicals

  - The expression $\sqrt[n]{x}$ denotes the "nth root of x," where x is the *radicand* and n is the *index*
  - When computing the square root of 16, denoted by $\sqrt{16}$, there are two numbers, -4 and 4, whose square is 16, since $4^2 = 16$ and $(-4)^2 = 16$. Convention holds that the positive square root, known as the *principal root*, is the value used.
  - Rules for radicals (shown for the index $n = 2$, but true for all n)

  $$\sqrt{ab} = \sqrt{a} \cdot \sqrt{b}$$
  $$\sqrt{\frac{a}{b}} = \frac{\sqrt{a}}{\sqrt{b}}$$

  - Radicals can be expressed as fractional exponents, for example,

  $$\sqrt{x} = x^{1/2} \qquad \sqrt[3]{x} = x^{1/3} \qquad \sqrt[n]{x} = x^{1/n}$$

  - Using the exponent rules

  $$x^{a/b} = (x^a)^{1/b} = \sqrt[b]{x^a}$$

  or

  $$x^{a/b} = \left(x^{1/b}\right)^a = \left(\sqrt[b]{x}\right)^a$$

  - Simplifying radicals
    * Perfect squares, cubes, etc., should be extracted when possible, for example,

    $$\sqrt{32} = \sqrt{16 \cdot 2} = \sqrt{16} \cdot \sqrt{2} = 4\sqrt{2}$$
    $$\sqrt[3]{-54} = \sqrt[3]{-27 \cdot 2} = \sqrt[3]{-27} \cdot \sqrt[3]{2} = -3\sqrt[3]{2}$$

* If the index and the power of the radicand share a common factor, the radical should be reduced, for example,
$$\sqrt[9]{x^6} = x^{6/9} = x^{2/3} = \sqrt[3]{x^2}$$
* If the power of the radicand exceeds the index, the radical should be reduced, for example,
$$\sqrt{x^5} = \sqrt{x^4 \cdot x} = \sqrt{x^4} \cdot \sqrt{x} = x^2\sqrt{x}$$
* Radicals in denominators can be eliminated by "rationalizing the denominator," for example,
$$\sqrt{\frac{2}{7}} = \frac{\sqrt{2}}{\sqrt{7}} = \frac{\sqrt{2} \cdot \sqrt{7}}{\sqrt{7} \cdot \sqrt{7}} = \frac{\sqrt{14}}{7}$$
* Radicals with the same index and radicand can be combined, for example,
$$\sqrt{8} - 5\sqrt{2} = 2\sqrt{2} - 5\sqrt{2} = -3\sqrt{2}$$

- Logarithms
  - The base b logarithm of a (or "log base b of a") is written as $\log_b a$, where $b > 0$ and $b \neq 1$
  - $\log_b a = c$ is equivalent to $b^c = a$; therefore

  $$\begin{array}{lll} \log_{10} 1000 = 3 & \text{is equivalent to} & 10^3 = 1000 \\ \log_{17} 1 = 0 & \text{is equivalent to} & 17^0 = 1 \\ \log_5(1/25) = -2 & \text{is equivalent to} & 5^{-2} = 1/25 \end{array}$$

  - The *natural logarithm* of x, $\ln x$, is shorthand for $\log_e x$ or "the logarithm base e of x"
  - Another popular base for logarithms is base 10, which is known as the *common logarithm* and is written as $\log_{10} x$, or simply $\log x$
  - For any base b,
  $$\log_b x = \log_b y \quad \Rightarrow \quad x = y$$
  - Rules for logarithms (for any fixed base)
  $$\log 1 = 0$$
  $$\log(xy) = \log x + \log y$$
  $$\log\left(\frac{x}{y}\right) = \log x - \log y$$
  $$\log x^a = a \log x$$
  - Logarithm identities (for any fixed base b)
  $$\log_b b = 1$$
  $$b^{\log_b x} = x$$
  $$\log_b(b^x) = x$$

- Simplifying logarithmic expressions
    * Logarithmic expressions can be expanded, for example,
    $$\ln\left(\frac{x^2 y}{\sqrt{z}}\right) = 2\ln x + \ln y - \frac{1}{2}\ln z$$
    * Logarithmic expressions can be contracted so they can be written as a single logarithm, for example,
    $$\frac{1}{3}\log x - \log y - \frac{1}{2}\log z = \log\left(\frac{x^{1/3}}{yz^{1/2}}\right)$$
- Computing logarithms on a calculator
    * If the base is 10, use the LOG key
    * If the base is e, use the LN key
    * For all other bases, either key the base into the calculator (if it has that capability) or use one of the *change of base formulas*:
    $$\log_b x = \frac{\log x}{\log b} \qquad \text{or} \qquad \log_b x = \frac{\ln x}{\ln b}$$

- Algebraic expressions
    - A *constant* is a real number, for example, 7, π, 2/3, 1.7, e^2, $\sqrt{17}$
    - A *variable* is a placeholder that may take on several values, for example, x, y, a, b
    - There are no hard and fast rules about the letter choices for constants and variables, but here are some patterns that have emerged over the years:
        * a, b, c are often constants or coefficients
        * d is often reserved for calculus, where it has a special meaning
        * e is often reserved for Euler's number $e = 2.718281828459045\ldots$
        * f, g, h are often function names, for example, $f(x)$
        * i, j, k, l, m, n are often integers or indexes, for example, $\sum_{i=1}^{n} i^2$
        * o is often used for special functions, for example, $o(h)$
        * p, q, r are often probabilities
        * s, t, u, v, w, x, y, z are general-purpose variables
        * Greek letters such as $\alpha, \beta, \gamma, \delta$ are often used to denote angles in geometry and trigonometry, for example, $\sin\theta$
        * Greek letters are also used to denote parameters in probability distributions, for example, $X \sim N(\mu, \sigma^2)$
    - The set of values for which a function (defined in Chapter 4) is defined is the *domain*, for example, the domain of $1/(x-3)$ is $\{x \mid x \in \mathbb{R}, x \neq 3\}$. The bar in the expression for the domain is read "such that."

Chapter 2. Algebra

- A *term* (a product of constants and variables) in an algebraic expression typically has the constant written before the variable, for example, $7x$, $13xy$, $\sqrt{3}z^3$, $\pi \log w$
- The constant factor in a term is known as the *coefficient*
- An algebraic expression involving one term is a *monomial*, for example, $7x$, $\frac{7}{8}t$, $16y^3$
- An algebraic expression involving two terms is a *binomial*, for example, $3x+8$, $x-2y^3$
- An algebraic expression involving three terms is a *trinomial*, for example, $3x^4 - 7x + 8$, $\log 3 - x - 2y^3$
- Three common operations that can be applied to algebraic expressions are *simplifying*, *expanding*, and *factoring*

- Simplifying algebraic expressions
  - Terms with identical variable portions should be combined, for example,
  $$7xy^2 - 3xy^2 + z = 4xy^2 + z$$
  - For the product of monomials, combine like variables using exponent rules, for example,
  $$(2x) \cdot (3xy^3) \cdot (4xz) = 24x^3y^3z$$

- Expanding algebraic expressions
  - F.O.I.L. (First, Outside, Inside, Last) is used for multiplying binomials, for example,
  $$(2x+3)(4x-5) = \underbrace{8x^2}_{\text{First}} - \underbrace{10x}_{\text{Outside}} + \underbrace{12x}_{\text{Inside}} - \underbrace{15}_{\text{Last}} = 8x^2 + 2x - 15$$
  - When multiplying two expressions, each with more than two terms, each term in the first expression must be multiplied by each term in the second expression and the resulting products are summed, for example,
  $$\begin{aligned}(x^3 + 6x - 5)(x^2 + 3x - 2) &= x^3(x^2 + 3x - 2) + 6x(x^2 + 3x - 2) - 5(x^2 + 3x - 2) \\ &= x^5 + 3x^4 - 2x^3 + 6x^3 + 18x^2 - 12x - 5x^2 - 15x + 10 \\ &= x^5 + 3x^4 + 4x^3 + 13x^2 - 27x + 10\end{aligned}$$

  This method works for multiplying binomials, such as $(2x+3)(4x-5)$, as well.
  - Well-known expansions
  $$\begin{aligned}(x+y)^2 &= x^2 + 2xy + y^2 \\ (x-y)^2 &= x^2 - 2xy + y^2 \\ (x+y)^3 &= x^3 + 3x^2y + 3xy^2 + y^3 \\ (x-y)^3 &= x^3 - 3x^2y + 3xy^2 - y^3\end{aligned}$$

- Binomial expansion

$$(x+y)^n = \sum_{i=0}^{n} \binom{n}{i} x^{n-i} y^i$$

  for example,

$$(x+y)^4 = x^4 + 4x^3y + 6x^2y^2 + 4xy^3 + y^4$$

- Factoring algebraic expressions

  - Well-known factoring formulas

$$\begin{aligned} x^2 - y^2 &= (x+y)(x-y) \\ x^2 + 2xy + y^2 &= (x+y)^2 \\ x^2 - 2xy + y^2 &= (x-y)^2 \\ x^3 - y^3 &= (x-y)\left(x^2 + xy + y^2\right) \\ x^3 + y^3 &= (x+y)\left(x^2 - xy + y^2\right) \end{aligned}$$

- Equations

  - An *equation* is a mathematical statement that two algebraic expressions are equal, for example, $x + 4 = 9$
  - To *solve* an equation is to perform mathematical operations in order to find the value(s) of the unknown variable that make the equation true
  - The value(s) that solve an equation are known as the *roots* or *solutions*
  - The set of all such solutions to an equation is the *solution set*
  - Values in the solution set are said to *satisfy* the equation
  - Examples:
    * Solve $x + 4 = 9$ for x. Solution: $x = 5$
    * Solve $F = ma$ for m. Solution: $m = F/a$
    * Solve $y^2 = 81$ for y. Solution set: $\{-9, 9\}$
  - Classifying solution sets
    * An equation whose solution set is *all* the real numbers is known as an *identity equation*, for example, $x + 2x = 3x$ has solution set $\mathbb{R}$
    * An equation whose solution set is *some* of the real numbers is known as a *conditional equation*, for example, $x^2 = 9$ has solution set $\{-3, 3\}$
    * An equation whose solution set is *empty* is known as an *inconsistent equation*, for example, $x = x + 1$ has an empty solution set

Chapter 2. Algebra

- Techniques for solving equations
  * Simplify one side of an equation by applying any of the basic arithmetic operations (such as $+$, $-$, $\times$, $\div$) to it, for example,
  $$x = (2+4)(6-3)$$
  $$x = 18$$
  * Apply any of the rules for exponents, radicals, and logarithms to one side of an equation, for example,
  $$x = \log\left(y^2 z^4\right)$$
  $$x = 2\log\left(yz^2\right)$$
  * Add or subtract a constant on both sides of an equation, for example,
  $$x + 2 = 8$$
  $$x + 2 - 2 = 8 - 2$$
  $$x = 6$$
  * Multiply or divide both sides of an equation by a nonzero constant, for example,
  $$3x = 24$$
  $$\frac{1}{3} \cdot 3x = \frac{1}{3} \cdot 24$$
  $$x = 8$$
  * Raise both sides of an equation to a nonzero constant power, for example,
  $$x^{1/3} = 2$$
  $$\left(x^{1/3}\right)^3 = 2^3$$
  $$x = 8$$
  * Use both sides of an equation as powers of the same base, for example,
  $$\ln x = 2$$
  $$e^{\ln x} = e^2$$
  $$x = e^2$$
  * Take the logarithm of both sides of an equation, for example,
  $$10^x = 20$$
  $$\log_{10}(10^x) = \log_{10}(20)$$
  $$x \log_{10}(10) = \log_{10}(20)$$
  $$x = \log_{10}(20)$$

* Switch the two sides of an equation, for example,
$$9 = x$$
$$x = 9$$

- The product of two or more expressions being equal to zero means at least one of the expressions must equal zero, for example,
$$x(x+4)(x-3)(x+\pi) = 0$$
implies that
$$x = 0 \quad \text{or} \quad x+4 = 0 \quad \text{or} \quad x-3 = 0 \quad \text{or} \quad x+\pi = 0$$
or, equivalently
$$x = 0 \quad \text{or} \quad x = -4 \quad \text{or} \quad x = 3 \quad \text{or} \quad x = -\pi$$

- Equations involving absolute value typically assume that the expression within the absolute value bars can assume two values, for example,
$$|17 - 4x| = 5$$
can be broken into
$$17 - 4x = 5 \quad \text{or} \quad 17 - 4x = -5$$
which can be solved individually, yielding the solutions $x = 3$ and $x = 11/2$

- *Extraneous roots* can be introduced using standard techniques for solving an equation, but are not solutions of the original equation. For example, the first step in solving
$$\sqrt{9 - x} = x + 3$$
for x is to square both sides of the equation to eliminate the radical, yielding
$$9 - x = (x+3)^2$$
or
$$9 - x = x^2 + 6x + 9$$
or
$$x^2 + 7x = 0$$
which can be factored as
$$x(x+7) = 0$$
which has roots $x = 0$ and $x = -7$. The first root solves the original equation because $\sqrt{9-0} = \sqrt{9} = 3$, but the second root is extraneous because
$$\sqrt{9 - (-7)} = \sqrt{16} = 4 \neq -4$$
The extraneous root was introduced when both sides of the equation were squared.

Chapter 2. Algebra

- A *system of equations* is two or more equations whose solution must satisfy all equations simultaneously, for example, the solution $x = 2$ and $y = 1$ solves the system of equations

$$2x - 3y = 1$$
$$3x + 2y = 8$$

One way to solve a system of two equations in two variables is to solve for one of the variables in the first equation, then substitute into the second equation. This gives an equation in one variable, which can be solved. The variable that has been solved for can then be plugged into either of the original equations to give an equation in one variable that can also be solved to find the value of the other variable.

- Polynomials
  - A degree n polynomial in the variable x written in standard form is

  $$a_n x^n + a_{n-1} x^{n-1} + \cdots + a_2 x^2 + a_1 x + a_0$$

  where the coefficients $a_n, a_{n-1}, \ldots, a_2, a_1, a_0$ are real constants, $a_n \neq 0$ is the *leading coefficient*, and a_0 is the *constant term*
  - Basic arithmetic operations
    * Addition

    $$\left(5x^3 - 2x^2 + 7x - 8\right) + \left(2x^3 - 14x + 3\right) = 7x^3 - 2x^2 - 7x - 5$$

    * Subtraction

    $$\left(3x^4 - 3x - 5\right) - \left(x^3 - 4x\right) = 3x^4 - x^3 + x - 5$$

    * Multiplication (polynomials of degree $n = 2$ via F.O.I.L.)

    $$(6x - 5)(3x + 4) = \underbrace{18x^2}_{\text{First}} + \underbrace{24x}_{\text{Outside}} - \underbrace{15x}_{\text{Inside}} - \underbrace{20}_{\text{Last}} = 18x^2 + 9x - 20$$

    * Multiplication (generally)

    $$\begin{aligned}(6x - 5)\left(x^2 + 3x - 2\right) &= 6x\left(x^2 + 3x - 2\right) - 5\left(x^2 + 3x - 2\right) \\ &= 6x^3 + 18x^2 - 12x - 5x^2 - 15x + 10 \\ &= 6x^3 + 13x^2 - 27x + 10\end{aligned}$$

    * Division

    $$\frac{2x^2 + 4}{6x^7 - 8x + 10} = \frac{2\left(x^2 + 2\right)}{2\left(3x^7 - 4x + 5\right)} = \frac{x^2 + 2}{3x^7 - 4x + 5}$$

  - Classifying polynomials
    * When $n = 1$, the polynomial is *linear*, or *first degree*, for example,

    $$2x + 9$$

* When $n = 2$, the polynomial is *quadratic*, or *second degree*, for example,
$$5x^2 - 6x + 4$$
* When $n = 3$, the polynomial is *cubic*, or *third degree*, for example,
$$x^3$$
* When $n = 4$, the polynomial is *quartic*, or *fourth degree*, for example,
$$-x^4 + 6x + 3/2$$
* When $n = 5$, the polynomial is *quintic*, or *fifth degree*, for example,
$$4x^5 + \pi x^3 - 17.3x^2 - e$$

- Techniques for factoring polynomials
    * Factoring by finding a common factor, for example,
    $$4x^5 - 8x^3 + 12x^2 = 4x^2\left(x^3 - 2x + 3\right)$$
    * Factoring polynomials that are differences of squares, for example,
    $$16x^2 - 49 = (4x+7)(4x-7)$$
    * Factoring quadratic polynomials as products of binomials (which is the inverse of F.O.I.L.), for example,
    $$x^2 - 2x - 8 = (x-4)(x+2)$$
    * Factoring by grouping, for example,
    $$\begin{aligned} x^6 - 3x^5 + 5x - 15 &= \left(x^6 - 3x^5\right) + (5x - 15) \\ &= x^5(x-3) + 5(x-3) \\ &= \left(x^5 + 5\right)(x-3) \end{aligned}$$

- Rational expressions
    * A *rational expression* is the ratio of two polynomials, for example,
    $$\frac{3x^2}{x-4}$$
    * The *domain* of a rational function is the set of all real numbers except those that yield a denominator of 0, for example, the domain of
    $$\frac{3x^2}{x-4}$$
    is $\{x \mid x \in \mathbb{R} \text{ and } x \neq 4\}$

Chapter 2. Algebra

- A degree n *polynomial equation* in the variable x written in standard form is

$$a_n x^n + a_{n-1} x^{n-1} + \cdots + a_2 x^2 + a_1 x + a_0 = 0$$

where the coefficients a_n, a_{n-1}, ..., a_2, a_1, a_0 are real constants and $a_n \neq 0$

- *Long division* of polynomials is similar to long division of integers and yields a result that can be stated in the general form

$$\frac{\text{dividend}}{\text{divisor}} = \text{quotient} + \frac{\text{remainder}}{\text{divisor}}$$

for example, $x^3 - 7x^2 - x + 9$ divided by $x - 1$ is computed via long division as

$$\begin{array}{r}
x^2 - 6x - 7 \\
x-1 \overline{\smash{)}\, x^3 - 7x^2 - x + 9} \\
\underline{-x^3 + x^2 } \\
-6x^2 - x \\
\underline{6x^2 - 6x } \\
-7x + 9 \\
\underline{7x - 7} \\
2
\end{array}$$

and can be expressed as

$$\frac{x^3 - 7x^2 - x + 9}{x - 1} = x^2 - 6x - 7 + \frac{2}{x - 1}$$

- *Synthetic division* of polynomials can be used when the divisor has the form $x - a$ for some constant value a, for example, $x^3 - 7x^2 - x + 9$ divided by $x - 1$ is computed via synthetic division as

$$\begin{array}{c|cccc}
 & 1 & -7 & -1 & 9 \\
1 & & 1 & -6 & -7 \\
\hline
 & 1 & -6 & -7 & 2
\end{array}$$

The algorithm for synthetic division follows.

* Draw horizontal and vertical lines as illustrated in the example
* Place the coefficients of the polynomial in the numerator in the first row, including zeros for missing powers of x
* Place a, the value associated with the denominator (divisor), in the second row to the left of the vertical line
* The leading coefficient of the numerator is copied directly below to the bottom row of integers unchanged

* Moving from left to right, execute the follow steps repeatedly until the third row has been filled in:
  · The number just written in the bottom row is multiplied by a and placed in the second row, one column to the right of the number just written
  · Add the number just written in the second row to the number above it in the first row, and write the sum directly below in the third row
* After the third row has been completed, its rightmost entry is the remainder, and the entries to the left of it are the coefficients of the quotient

- The *fundamental theorem of algebra*, which states that a degree n polynomial equation has at least one root (real or complex), implies that a degree n polynomial equation has exactly n roots (also called solutions) when each root is counted up to its multiplicity. Not all of the roots are necessarily distinct and some of the roots may be real numbers and others may be complex numbers. For example, the third degree polynomial ($n = 3$)

$$x^3 - x^2 - 8x + 12$$

can be factored as

$$x^3 - x^2 - 8x + 12 = (x-2)^2(x+3)$$

so that the associated polynomial equation

$$x^3 - x^2 - 8x + 12 = (x-2)^2(x+3) = 0$$

has roots 2 (multiplicity 2) and -3 (multiplicity 1). So the number of roots is $2+1 = 3$, which is the degree of the polynomial. Complex numbers will be introduced in Chapter 6.

- A *linear equation* (a degree 1 polynomial equation) in x of the form

$$ax + b = 0$$

where $a \neq 0$, can be solved by subtracting b from both sides, then dividing both sides by a, for example,

$$2x + 8 = 0$$
$$2x = -8$$
$$x = -4$$

- The *quadratic equation*

$$x^2 = 25$$

has two solutions: $x = -5$ and $x = 5$, even though $\sqrt{25} = 5$ (by definition)

- A *quadratic equation* (a degree 2 polynomial equation) in x of the form

$$ax^2 + bx + c = 0$$

can be solved by three methods:

Chapter 2. Algebra

* factoring, for example,

$$x^2 + 2x - 8 = 0 \quad \Rightarrow \quad (x+4)(x-2) = 0 \quad \Rightarrow$$
$$x+4 = 0 \text{ or } x-2 = 0 \quad \Rightarrow \quad x = -4 \text{ or } x = 2$$

* the quadratic formula

$$x = \frac{-b \pm \sqrt{b^2 - 4ac}}{2a}$$

for example,

$$x^2 + 2x - 8 = 0 \quad \Rightarrow \quad a = 1, b = 2, c = -8 \quad \Rightarrow$$
$$x = \frac{-2 \pm \sqrt{4 + 32}}{2} \quad \Rightarrow \quad x = \frac{-2 \pm 6}{2} \quad \Rightarrow \quad x = -4 \text{ or } x = 2$$

* completing the square, for example,

$$x^2 + 2x - 8 = 0 \quad \Rightarrow \quad \left(x^2 + 2x + 1\right) - 1 - 8 = 0 \quad \Rightarrow$$
$$(x+1)^2 = 9 \quad \Rightarrow \quad x + 1 = \pm 3 \quad \Rightarrow \quad x = -1 \pm 3 \quad \Rightarrow \quad x = -4 \text{ or } x = 2$$

- Factoring a polynomial when a root is known

  If a is a root of a polynomial, then $(x - a)$ is a factor of that polynomial. For example, $x = 2$ is a root of the polynomial $x^3 - 7x + 6$ because $2^3 - 7 \cdot 2 + 6 = 8 - 14 + 6 = 0$, so $(x - 2)$ is a factor of $x^3 - 7x + 6$. Subsequent long division or synthetic division shows that $x^3 - 7x + 6 = (x - 2)\left(x^2 + 2x - 3\right)$.

- Inherently quadratic equations

  * Higher-order polynomial equations

  $$x^6 - 3x^3 + 2 = 0$$
  $$\left(x^3\right)^2 - 3\left(x^3\right) + 2 = 0$$
  $$\left(x^3 - 2\right)\left(x^3 - 1\right) = 0$$
  $$x^3 - 2 = 0 \quad \text{or} \quad x^3 - 1 = 0$$
  $$x^3 = 2 \quad \text{or} \quad x^3 = 1$$
  $$x = \sqrt[3]{2} \quad \text{or} \quad x = 1$$

  * Expressions involving radicals

  $$x^{2/5} - 3x^{1/5} + 2 = 0$$
  $$\left(x^{1/5}\right)^2 - 3\left(x^{1/5}\right) + 2 = 0$$
  $$\left(x^{1/5} - 2\right)\left(x^{1/5} - 1\right) = 0$$
  $$x^{1/5} - 2 = 0 \quad \text{or} \quad x^{1/5} - 1 = 0$$
  $$x^{1/5} = 2 \quad \text{or} \quad x^{1/5} = 1$$
  $$x = 32 \quad \text{or} \quad x = 1$$

- Inequalities
    - Inequalities indicate the ordering of quantities on a number line
    - The expression $2 < 5$ is read "two is less than five" and indicates that 2 lies to the left of 5 on a number line
    - Inequalities are mathematical statements of the form

$x < y$	x is less than y
$x \leq y$	x is less than or equal to y
$x > y$	x is greater than y
$x \geq y$	x is greater than or equal to y

    - The inequalities $x < y$ and $x > y$ are known as *strict inequalities*
    - Interval notation
        * The set of x values satisfying $a \leq x \leq b$ is a *closed interval* and is denoted by $[a, b]$

            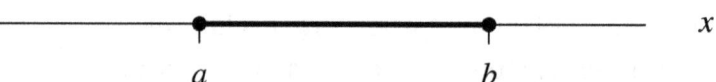

        * The set of x values satisfying $a < x < b$ is an *open interval* and is denoted by (a, b)

            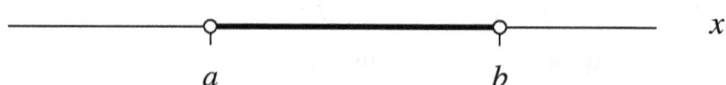

        * The set of x values satisfying $a \leq x < b$ is denoted by $[a, b)$

        * The set of x values satisfying $a < x \leq b$ is denoted by $(a, b]$

        * The set of x values satisfying $x \leq b$ is denoted by $(-\infty, b]$

            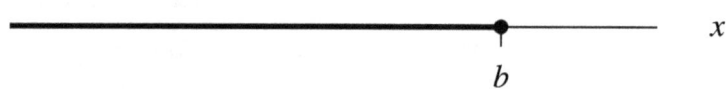

        * The set of x values satisfying $-\infty < x < \infty$ is the entire real number line and is denoted by $(-\infty, \infty)$
    - For a positive constant a, $|x| = a$ is equivalent to $x = \pm a$
    - For a positive constant a, $|x| < a$ is equivalent to $-a < x < a$

Chapter 2. Algebra

- For a positive constant a, $|x| > a$ is equivalent to $x < -a$ or $x > a$
- Multiplying or dividing an inequality by a negative number reverses the sense of the inequality, for example, multiplying by -1,

$$-x > 3 \quad \Rightarrow \quad x < -3$$

- *Inequalities involving linear expressions* are solved by adding constants to all expressions in an inequality and by multiplying all expressions by constants, for example,

$$5 < 2x - 7 < 23$$

$$12 < 2x < 30$$

$$6 < x < 15$$

so the solution set is

$$\{x \mid x \in \mathbb{R} \text{ and } 6 < x < 15\}$$

which is the open interval $(6, 15)$ as illustrated below.

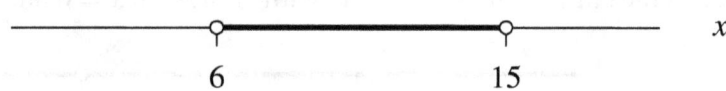

- *Inequalities involving quadratic expressions* are solved by finding the solutions to the associated quadratic equation, then testing points on both sides of each solution in the inequality. If a test point on an interval satisfies the inequality, then the entire interval satisfies the inequality. For example,

$$x^2 - 3x + 2 > 0$$

has associated quadratic equation

$$x^2 - 3x + 2 = 0$$

which can be factored as

$$(x-1)(x-2) = 0$$

with solutions $x = 1$ and $x = 2$. Checking the test points $x = 0$, $x = 3/2$, and $x = 3$ in the original inequality yields the solution set

$$\{x \mid x \in \mathbb{R} \text{ and } x < 1 \text{ or } x > 2\}$$

as illustrated below.

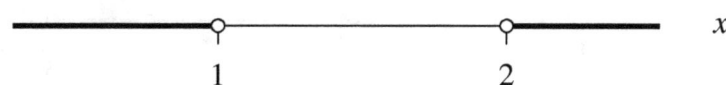

- *Inequalities involving polynomial expressions* are solved by finding the solutions to the associated polynomial equation, then testing points on both sides of each solution in the inequality. If a test point on an interval satisfies the inequality, then the entire interval satisfies the inequality. For example,

$$x^2 \left(x^2 - 6x + 9\right) > 0$$

has associated polynomial equation

$$x^2 \left(x^2 - 6x + 9\right) = 0$$

which can be factored as

$$x^2(x-3)^2 = 0$$

with solutions $x = 0$ and $x = 3$. Checking the test points $x = -1$, $x = 2$, and $x = 4$ in the original inequality yields the solution set

$$\{x \,|\, x \in \mathbb{R} \text{ and } x \neq 0 \text{ and } x \neq 3\}$$

which is the entire real number line with "holes" at $x = 0$ and $x = 3$ as illustrated below.

- *Inequalities involving rational expressions* are solved by finding the solutions to the associated polynomial equations in both the numerator and denominator, then testing points on both sides of each of these solutions in the inequality. For example, the first step in solving the inequality

$$\frac{x-2}{x^2-1} \leq 0$$

is to find the roots to the equations $x - 2 = 0$ and $x^2 - 1 = 0$. These roots are $x = -1$, $x = 1$, and $x = 2$. The roots $x = -1$ and $x = 1$ involve division by zero in the rational expression, so they are not in the solution set. The root $x = 2$ satisfies the inequality, so it is in the solution set. The next step in solving the inequality is to try test points on either side of each of the three roots (there are four such test points). If a test point satisfies the inequality, then the entire interval satisfies the inequality. Using the test points $x = -2$, $x = 0$, $x = 3/2$, and $x = 3$, the solution set for the inequality is

$$\{x \,|\, x \in \mathbb{R} \text{ and } x < -1 \text{ or } 1 < x \leq 2\}$$

as illustrated below.

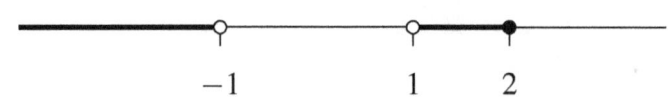

Chapter 2. Algebra

- Word problems

  - General steps for solving word problems
    1. Read and understand the problem and determine the unknown(s) of interest
    2. Choose an appropriate variable for each unknown, for example, let v be velocity
    3. Relate variables via equations
    4. Solve the equations for the variable of interest
    5. If appropriate, exclude any extraneous solutions
    6. Check the solution
    7. Translate the solution back to English

  - Sample problem using the steps listed above: Find three consecutive positive odd integers whose sum is 4 less than the square of the smallest integer.
    1. The unknowns of interest are the three consecutive positive odd integers
    2. Let n be the smallest positive odd integer; let $n+2$ be the middle positive odd integer; let $n+4$ be the largest positive odd integer
    3. Relate variables via equations:

Text	Mathematics
Find three consecutive positive odd integers	$n, n+2, n+4$
whose sum	$n+(n+2)+(n+4)$
is	$=$
4 less than the square of the smallest integer.	$n^2 - 4$

        So the equation of interest is

        $$n+(n+2)+(n+4) = n^2 - 4$$

    4. Solve the equations for the variable of interest

        $$3n + 6 = n^2 - 4$$
        $$n^2 - 3n - 10 = 0$$
        $$(n+2)(n-5) = 0$$
        $$n = -2 \quad \text{or} \quad n = 5$$

    5. The two sequences of three integers: $-2, 0, 2$ and $5, 7, 9$, satisfy the equation, but the first sequence of three is excluded because the integers are even and the smallest integer is negative
    6. Check the solution. The sum of the three integers must be 4 less than the square of the smallest integer:
        $$5 + 7 + 9 = 5^2 - 4$$
        $$21 = 21$$

7. Translate the solution back to English. The three consecutive positive odd integers are 5, 7, and 9.

- Special classes of word problems
    * *Distance and travel problems* typically employ the formula
    $$d = rt$$
    where d is distance, r is rate, and t is time. It is important to keep the units consistent in distance and travel problems. For example, if the rate is given in miles per hour, then the distance must be in miles and the time must be in hours. This equation can be rearranged to give
    $$t = \frac{d}{r} \quad \text{and} \quad r = \frac{d}{t}$$
    so it is not necessary to remember all three equations.
    * *Work problems* typically involve several people working together on a project
    * *Mixture problems* typically involve mixing solutions or substances having various concentrations
    * *Interest problems* involve money accumulating in an account
        · Simple interest:
        $$I = Prt$$
        where I is the interest, P is the principal (original investment), r is the annual interest rate as a decimal (not as a percentage), and t is time in years
        · Compound interest (periodic compounding):
        $$A = P\left(1 + \frac{r}{n}\right)^{nt}$$
        where A is the value at time t, P is the principal (original investment), r is the annual interest rate as a decimal (not as a percentage), n is the number of times the interest is compounded annually, and t is time in years
        · Compound interest (continuous compounding):
        $$A = Pe^{rt}$$
        where A is the value at time t, P is the principal (original investment), r is the annual interest rate as a decimal (not as a percentage), and t is time in years

Exercises

2.1 Find $\log_7 49$.

2.2 Simplify $\log_3 18 - \log_3 2$.

2.3 Solve $3x - 8 = 19$ for x.

2.4 Solve $2^x 4 = 8^{22}$ for x.

2.1 Find $\log_7 49$.

Letting the desired quantity be written as $x = \log_7 49$, this equation can be written in exponential form as $7^x = 49$. So $x = 2$ and thus

$$\log_7 49 = 2.$$

2.2 Simplify $\log_3 18 - \log_3 2$.

Using the logarithm rules

$$\log_3 18 - \log_3 2 = \log_3 \frac{18}{2} = \log_3 9 = 2.$$

2.3 Solve $3x - 8 = 19$ for x.

Adding 8 to both sides of the equation gives

$$3x = 27.$$

Dividing both sides of this equation by 3 gives the solution

$$x = 9.$$

2.4 Solve $2^x 4 = 8^{22}$ for x.

Using the common base of 2, this equation can be rewritten as

$$2^x 2^2 = (2^3)^{22}$$

or

$$2^{x+2} = 2^{66}.$$

Equating the exponents gives

$$x + 2 = 66.$$

Subtracting 2 from both sides of this equation gives

$$x = 64.$$

Chapter 2. Algebra

2.5 Solve
$$\log_8 x = \frac{4}{3}$$
for x.

2.6 Solve
$$\log_b \frac{1}{32} = -5$$
for b.

2.7 Solve
$$\log_5(x-3) = 2$$
for x.

2.8 Write the following expression as the logarithm of a single quantity:
$$\log x + \log\left(xy^2\right) - \log z.$$

2.5 Solve
$$\log_8 x = \frac{4}{3}$$
for x.

Converting the equation to its exponential form,
$$x = 8^{4/3} = \left(2^3\right)^{4/3} = 2^4 = 16.$$

2.6 Solve
$$\log_b \frac{1}{32} = -5$$
for b.

Converting the equation to its exponential form,
$$b^{-5} = \frac{1}{32}$$
or
$$b^{-5} = 2^{-5},$$
so, equating the bases, $b = 2$.

2.7 Solve
$$\log_5(x-3) = 2$$
for x.

Converting the function to exponential form
$$x - 3 = 5^2$$
or
$$x - 3 = 25.$$
Adding 3 to both sides of this equation gives
$$x = 28.$$

2.8 Write the following expression as the logarithm of a single quantity:
$$\log x + \log\left(xy^2\right) - \log z.$$

Using the laws for logarithms
$$\log x + \log\left(xy^2\right) - \log z = \log\left(\frac{x \cdot xy^2}{z}\right) = \log\left(\frac{x^2 y^2}{z}\right).$$

Chapter 2. Algebra

2.9 Use factoring to solve $2x^2 - 7x + 3 = 0$.

2.10 Use completing the square to solve $x^2 - 8x + 2 = 0$.

2.11 Find the sum of the two solutions of the equation
$$ax^2 + bx + c = 0$$
for real constants $a \neq 0$, b, and c.

2.9 Use factoring to solve $2x^2 - 7x + 3 = 0$.

The quadratic factors into the two terms
$$2x^2 - 7x + 3 = (2x - 1)(x - 3) = 0.$$

Since one of these factors must be zero, one the two equations
$$2x - 1 = 0 \qquad x - 3 = 0$$

must hold. Solving these two equations gives
$$x = \frac{1}{2} \qquad \text{or} \qquad x = 3.$$

2.10 Use completing the square to solve $x^2 - 8x + 2 = 0$.

First, move the constant term to the right-hand side of the equation:
$$x^2 - 8x = -2.$$

Now add the square of half the coefficient of the linear term to both sides of the equation:
$$x^2 - 8x + 16 = -2 + 16.$$

The left-hand side of the equation is now a perfect square:
$$(x - 4)^2 = 14.$$

Taking the square root of both sides of the equation gives
$$x - 4 = \pm\sqrt{14}.$$

Finally, adding 4 to both sides of the equation gives the solutions
$$x = 4 \pm \sqrt{14}.$$

2.11 Find the sum of the two solutions of the equation
$$ax^2 + bx + c = 0$$

for real constants $a \neq 0$, b, and c.

Using the quadratic formula, the two solutions of the equation are
$$x = \frac{-b + \sqrt{b^2 - 4ac}}{2a} \qquad \text{and} \qquad x = \frac{-b - \sqrt{b^2 - 4ac}}{2a}.$$

The sum of these values is
$$\frac{-b + \sqrt{b^2 - 4ac}}{2a} + \frac{-b - \sqrt{b^2 - 4ac}}{2a} = \frac{-2b}{2a} = -\frac{b}{a}.$$

Chapter 2. Algebra

2.12 Find all values of x that satisfy
$$x + \frac{3}{x} = 4.$$

2.13 Mark can mow the lawn in 20 minutes. Lindsey can mow the lawn in 30 minutes. How long will it take the two of them to mow the lawn if they work together?

2.14 Solve
$$A = P\left(1 + \frac{r}{n}\right)^{nt}$$
for t.

2.12 Find all values of x that satisfy
$$x + \frac{3}{x} = 4.$$

Multiplying both sides of this equation by x and moving all terms to the left-hand side of the equation gives
$$x^2 - 4x + 3 = 0.$$

(Multiplying by a variable expression can introduce false solutions.) This quadratic can be factored as
$$(x-1)(x-3) = 0$$
which yields $x = 1$ and $x = 3$, both of which satisfy the original equation.

2.13 Mark can mow the lawn in 20 minutes. Lindsey can mow the lawn in 30 minutes. How long will it take the two of them to mow the lawn if they work together?

Mark can mow $1/20$ of the lawn in one minute. Lindsey can mow $1/30$ of the lawn in one minute. Together they can mow $1/20 + 1/30 = 1/12$ of the lawn in one minute. Therefore, working together they can mow the lawn in 12 minutes.

2.14 Solve
$$A = P\left(1 + \frac{r}{n}\right)^{nt}$$
for t.

Divide both sides of the equation by P:
$$\frac{A}{P} = \left(1 + \frac{r}{n}\right)^{nt}.$$

Take the logarithm of both sides of this equation:
$$\log\left(\frac{A}{P}\right) = \log\left(1 + \frac{r}{n}\right)^{nt}.$$

Use one of the logarithm laws to bring the exponent in front of the logarithm:
$$\log\left(\frac{A}{P}\right) = nt \log\left(1 + \frac{r}{n}\right).$$

Finally, divide both sides of this equation by $n \log(1 + r/n)$:
$$t = \frac{\log(A/P)}{n \log(1 + r/n)}.$$

Chapter 2. Algebra

2.15 What annual rate of interest, compounded 12 times per year, is needed to grow an initial deposit of $1000 to $1200 in seven years?

2.16 Solve
$$2^{4x-7} = 32$$
for x.

2.15 What annual rate of interest, compounded 12 times per year, is needed to grow an initial deposit of $1000 to $1200 in seven years?

It is first necessary to solve
$$A = P\left(1 + \frac{r}{n}\right)^{nt}$$
for r. First, divide both sides of the equation by P:
$$\frac{A}{P} = \left(1 + \frac{r}{n}\right)^{nt}.$$
Next, raise both sides of this equation to the $1/nt$ power:
$$\left(\frac{A}{P}\right)^{1/nt} = 1 + \frac{r}{n}.$$
Next, subtract 1 from both sides of this equation:
$$\left(\frac{A}{P}\right)^{1/nt} - 1 = \frac{r}{n}.$$
Finally, multiply both sides of this equation by n:
$$r = n\left[\left(\frac{A}{P}\right)^{1/nt} - 1\right].$$
Inserting the parameters $n = 12$, $P = 1000$, $A = 1200$, and $t = 7$, this becomes
$$r = 12\left[\left(\frac{1200}{1000}\right)^{1/(12 \cdot 7)} - 1\right] = 12\left[\left(\frac{6}{5}\right)^{1/84} - 1\right] \cong 0.026.$$
The account must earn 2.6% interest with monthly compounding to grow from $1000 to $1200 in seven years.

2.16 Solve
$$2^{4x-7} = 32$$
for x.

Using the common base of 2, this equation can be written as
$$2^{4x-7} = 2^5.$$
Equating exponents,
$$4x - 7 = 5.$$
Adding 7 to each side of this equation,
$$4x = 12.$$
Dividing both sides of this equation by 4,
$$x = 3.$$

2.17 If $\log_b x = p$ and $\log_b y = q$, write
$$\log_b \left(\frac{x}{y}\right)^3$$
in terms of p and q in its simplest form.

2.18 If $8^x = 4$ and $27^{2x-3y} = 9$, find y.

2.17 If $\log_b x = p$ and $\log_b y = q$, write
$$\log_b \left(\frac{x}{y}\right)^3$$
in terms of p and q in its simplest form.

Using the laws of logarithms,
$$\log_b \left(\frac{x}{y}\right)^3 = 3\log_b \left(\frac{x}{y}\right) = 3\left(\log_b x - \log_b y\right) = 3\left(p - q\right).$$

2.18 If $8^x = 4$ and $27^{2x-3y} = 9$, find y.

The first equation contains x alone, so it can be solved for x. Using the common base of 2, it can be rewritten as
$$\left(2^3\right)^x = 2^2$$
or
$$2^{3x} = 2^2.$$
Equating exponents,
$$3x = 2.$$
Dividing both sides of this equation by 3 gives
$$x = 2/3.$$
The second equation can be rewritten using the common base of 3 as
$$\left(3^3\right)^{2x-3y} = 3^2$$
or
$$3^{6x-9y} = 3^2.$$
Equating exponents,
$$6x - 9y = 2.$$
Substituting $x = 2/3$ gives
$$4 - 9y = 2.$$
Subtracting 4 from both sides of this equation gives
$$-9y = -2.$$
Finally, dividing both sides of this equation by -9 results in
$$y = \frac{2}{9}.$$

Chapter 2. Algebra

2.19 Solve
$$4^{x^2-3} = 32$$
for x.

2.20 Compute the quotient
$$\frac{x^3 + x^2 - 9x + 7}{x - 1}.$$

2.19 Solve
$$4^{x^2-3} = 32$$
for x.

Using the common base of 2, this equation can be rewritten as
$$\left(2^2\right)^{x^2-3} = 2^5$$
or
$$2^{2x^2-6} = 2^5.$$
Equating exponents,
$$2x^2 - 6 = 5.$$
Adding six to both sides of this equation, then dividing by 2 gives
$$x^2 = \frac{11}{2}.$$
Finally, taking the square root of both sides of this equation gives
$$x = \pm\sqrt{\frac{11}{2}}.$$

2.20 Compute the quotient
$$\frac{x^3 + x^2 - 9x + 7}{x - 1}.$$

Using long division:

$$\begin{array}{r}
x^2 + 2x - 7 \\
x-1{\overline{\smash{\big)}\,x^3 + x^2 - 9x + 7}} \\
\underline{-x^3 + x^2} \\
2x^2 - 9x \\
\underline{-2x^2 + 2x} \\
-7x + 7 \\
\underline{7x - 7} \\
0
\end{array}$$

So the quotient is $x^2 + 2x - 7$. Synthetic division gives the same quotient with less writing:

$$\begin{array}{c|cccc}
 & 1 & 1 & -9 & 7 \\
1 & & 1 & 2 & -7 \\
\hline
 & 1 & 2 & -7 & 0
\end{array}$$

Chapter 2. Algebra

2.21 Solve
$$\log 3 + 2\log x = \log(1 - 2x)$$
for x.

2.22 Solve
$$x^2 + 2xy^2 + y = -27$$
$$x + y = 3$$
for x and y.

2.21 Solve
$$\log 3 + 2\log x = \log(1-2x)$$
for x.

Using the logarithm rules
$$\log 3 + \log\left(x^2\right) = \log(1-2x)$$
or
$$\log\left(3x^2\right) = \log(1-2x).$$
Equating the arguments of the logarithms,
$$3x^2 = 1 - 2x$$
or
$$3x^2 + 2x - 1 = 0.$$
This quadratic equation can be factored as
$$(3x-1)(x+1) = 0$$
which implies that
$$3x - 1 = 0 \qquad \text{or} \qquad x + 1 = 0$$
or, equivalently,
$$x = 1/3 \qquad \text{or} \qquad x = -1.$$
Because $x = -1$ can't be used in the original equation (the logarithm of a negative number is not defined), the solution is $x = 1/3$.

2.22 Solve
$$x^2 + 2xy^2 + y = -27$$
$$x + y = 3$$
for x and y.

The second equation can be solved for x as $x = 3 - y$. Inserting this into the first equation,
$$(3-y)^2 + 2(3-y)y^2 + y = -27$$
which is a cubic polynomial in y that simplifies to
$$2y^3 - 7y^2 + 5y - 36 = 0.$$

This equation can be solved using synthetic division to yield a single real root $y = 4$. Then $y = 4$ can be inserted into the second equation to give $x = -1$. So the solution to the system of equations is $x = -1$ and $y = 4$.

Chapter 2. Algebra

2.23 Find all solutions to the equation
$$e^{6x} - e^{3x} - 2 = 0.$$

2.24 Find the solution set of the inequality
$$x^2 - 2x - 3 < 0.$$

2.23 Find all solutions to the equation

$$e^{6x} - e^{3x} - 2 = 0.$$

This equation is quadratic in e^{3x}:

$$\left(e^{3x}\right)^2 - e^{3x} - 2 = 0.$$

This quadratic can be factored as

$$\left(e^{3x} - 2\right)\left(e^{3x} + 1\right) = 0.$$

This equation is satisfied when one or both of the factors is zero, or when

$$e^{3x} - 2 = 0 \quad \text{or} \quad e^{3x} + 1 = 0$$

or, equivalently,

$$e^{3x} = 2 \quad \text{or} \quad e^{3x} = -1.$$

The second equation has no solution because e^{3x} is positive for all values of x. To solve the first equation, take the natural logarithm of both sides:

$$\ln e^{3x} = \ln 2,$$

which simplifies to

$$3x = \ln 2.$$

Dividing both sides of this equation by 3 gives the solution:

$$x = \frac{\ln 2}{3} \cong 0.231.$$

The decimal solution (0.231) that can be computed using a calculator is approximate because the exact solution $x = (\ln 2)/3$ is irrational.

2.24 Find the solution set of the inequality

$$x^2 - 2x - 3 < 0.$$

Since

$$x^2 - 2x - 3 = (x+1)(x-3),$$

the solutions of the associated polynomial equation are $x = -1$ and $x = 3$. The inequality should be checked for any point in the three test regions $(-\infty, -1)$, $(-1, 3)$, and $(3, \infty)$. For example, the inequality fails for $x = -2$, succeeds for $x = 0$, and fails for $x = 4$. So the solution set for the inequality is

$$\{x \,|\, x \in \mathbb{R}, -1 < x < 3\}.$$

Chapter 3

Geometry

Geometry considers the shape, size, and position of objects such as line segments, circles, triangles, and cubes. Geometry is often divided into *plane geometry* and *solid geometry* corresponding to two and three dimensions. Measurements associated with these objects include length, area, perimeter, and volume. Some mathematical results concerning these objects (for example, the sum of the angles in any triangle is 180°) are surveyed here.

- Points

    - A *point* is a geometrical object that defines a position
    - Graphically, a point is usually indicated by a small dot
    - A point is a 0-dimensional object
    - A point has no length, area, or volume
    - Capital letters are often used to denote points, for example, the points P and Q

- Lines, line segments, and rays

    - A *line* is a geometrical object of no width that is straight and infinitely long

    - A *ray* is a line extending from a point in a particular direction

        P

    - A *line segment* is two points on a line and all of the points between them

        P Q

    - The two points defining a line segment are known as the *endpoints*
    - If P and Q are the endpoints of a line segment, the line segment is often denoted by $\overline{PQ}$

- A line segment has one dimension, length, which is the distance between the two endpoints
- Two lines are *parallel* if they do not intersect and are always the same distance apart (this implies that the two lines lie in the same plane)
- Two lines are *perpendicular*, denoted by ⊥, if they intersect to form right angles
- A set of points that lie on the same line are *collinear* points
- Lines, line segments, and rays contain an infinite number of points
- Lines, line segments, and rays are one-dimensional objects

- Planes

    - A *plane* is a surface such that a line segment joining any two points on the surface lies entirely on the surface
    - Intuitively, if this sheet of paper is extended out to infinity in all directions, it is a plane
    - A plane is a two-dimensional object

- Angles

    - An *angle* consists of two different (non-collinear) rays that share a common endpoint

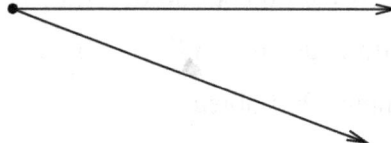

    - An angle is a two-dimensional object
    - The common endpoint of the two rays comprising an angle is known as the *vertex* of the angle
    - For an angle with vertex B, if the point A is on one of the rays and the point C is on the other ray, then $\angle ABC$ is used to denote the angle

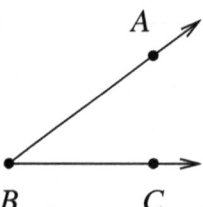

    - The *size* or *measure* of an angle is the smallest counterclockwise rotation about the vertex that moves one of the rays to the other
    - Degree measures for angles
        * Angles are often measured in *degrees*

Chapter 3. Geometry

* Degrees are designated by a circular superscript, for example, 47 degrees is denoted by 47°
* A full rotation is 360°

– Classifying angles
* An angle with a measure that is one fourth of a full rotation, that is, 90°, is a *right angle*
* Right angles are indicated by a brace near the vertex of the angle

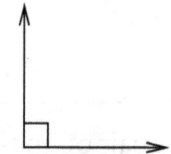

* An angle with degree measure between 0° and 90° is an *acute angle*

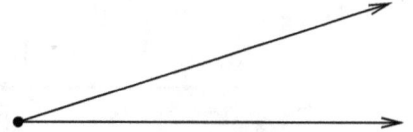

* An angle with degree measure between 90° and 180° is an *obtuse angle*

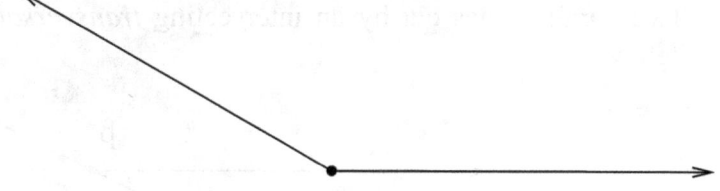

* An angle with degree measure 180° is a *straight angle*

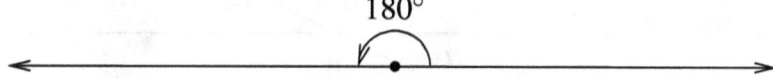

– Classifying pairs of angles (Greek letters denote angle measures)
* Two angles that have the same degree measure are *congruent angles*

* Two angles that have the same vertex and share a common ray are *adjacent angles*

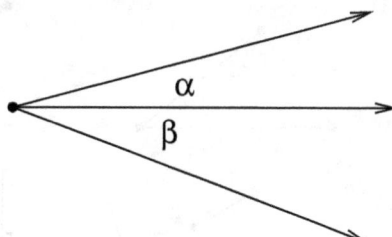

* Two angles whose degree sum is 90° are *complementary angles*

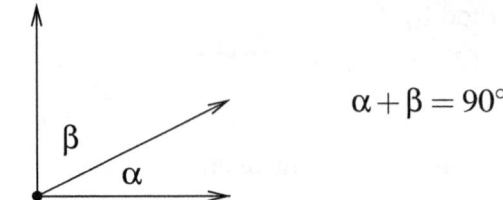

$$\alpha + \beta = 90°$$

* Two angles whose degree sum is 180° are *supplementary angles*

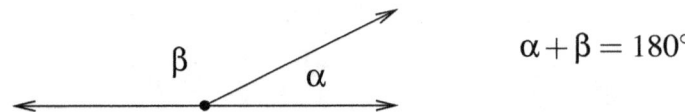

$$\alpha + \beta = 180°$$

* *Vertical angles* are a pair of non-adjacent angles formed by intersecting lines. Vertical angles have equal measures.

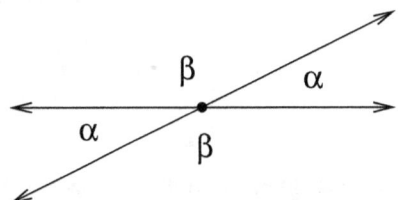

* Two parallel lines cut by an intersecting *transversal line* have measures shown below

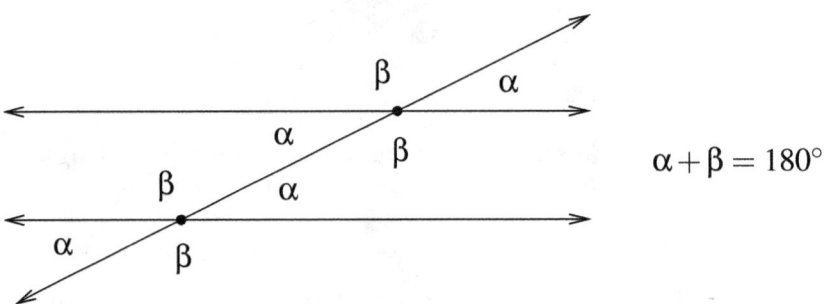

$$\alpha + \beta = 180°$$

- Polygons

  – A *polygon* consists of nonintersecting line segments joined together at their endpoints to form a two-dimensional closed figure with nonzero area

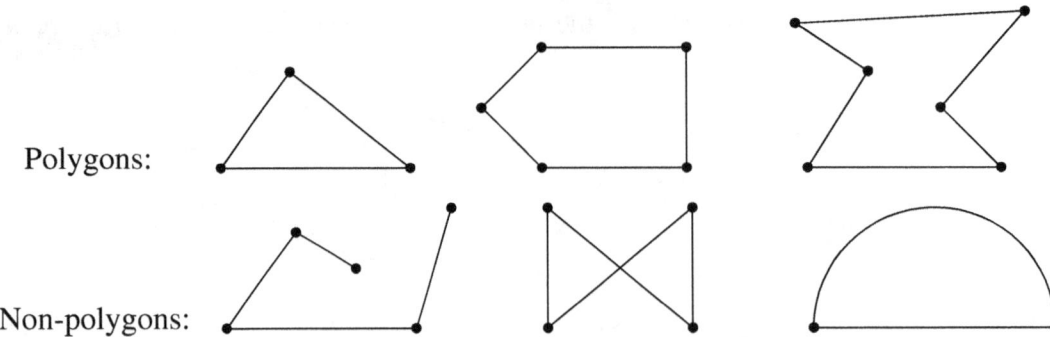

Chapter 3. Geometry

- The line segments comprising a polygon are known as the *sides* of the polygon
- The endpoints of the line segments comprising a polygon are known as the *vertices* of the polygon
- A polygon is a two-dimensional object
- The region enclosed by the sides of the polygon is known as the *interior* of the polygon
- The region outside of the polygon is known as the *exterior* of the polygon

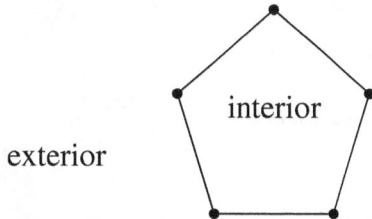

- An *n*-sided polygon has *n interior angles* measured into the interior of the polygon
- The degree sum of the measures of the interior angles of an *n*-sided polygon is

$$180(n-2)°$$

A triangle, for example, with $n = 3$ sides, has interior angles that sum to $180°$.

- An *n*-sided polygon has *n exterior angles* that are supplementary to the interior angles
- The degree sum of the measures of the exterior angles of an *n*-sided polygon is $360°$. This can be understood intuitively by considering a walk around an *n*-sided polygon that is marked on the floor. Each turn that is made at a vertex has a degree measure that corresponds to an exterior angle. One makes a full turn by walking around an *n*-sided polygon in this fashion, so the exterior angles must sum to $360°$.

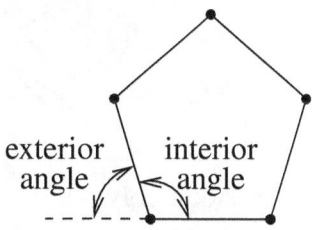

- Polygons are often classified by the number of sides

    * triangle: a three-sided polygon

    * quadrilateral: a four-sided polygon

    * pentagon: a five-sided polygon

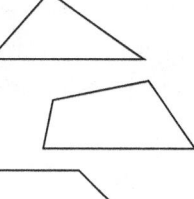

* hexagon: a six-sided polygon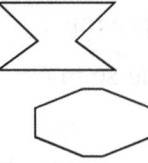
* octagon: an eight-sided polygon
- A *regular polygon* has sides of equal length and equal interior angles

Regular polygons: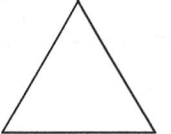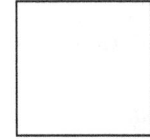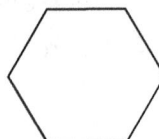

The area of a regular polygon is

$$A = \frac{1}{2}ap$$

where a is the *apothem* (the distance from the center of the regular polygon to a side) and p is the perimeter

- Convex and concave polygons
    * A polygon is *convex* if a line segment connecting any two points in the interior of the polygon lies completely within the polygon
    * A polygon is *concave* if it is not convex
    * Concave polygons have "dents"
    * Every interior angle of a convex polygon is at most 180°
    * A concave polygon has at least one interior angle that exceeds 180°

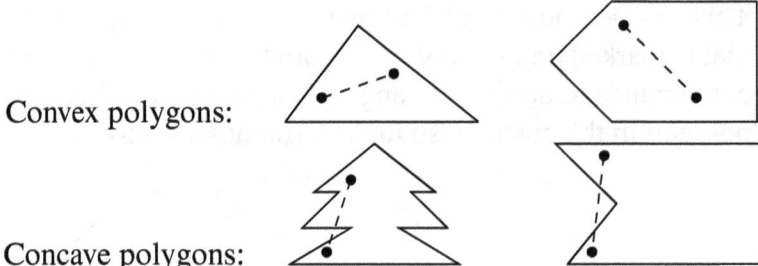

Convex polygons:

Concave polygons:

- Triangles
    * A *triangle* is a three-sided polygon
    * A triangle has three vertices
    * A triangle has three sides
    * A triangle has three angles whose degree measures sum to 180°
    * The measure of an exterior angle of a triangle is the sum of the measures of the two non-adjacent interior angles

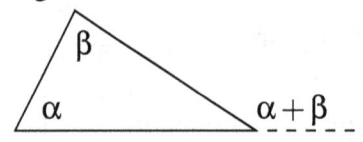

Chapter 3. Geometry

* All triangles are convex
* Classifying triangles
    · Two triangles are *similar* if their corresponding sides are proportional

    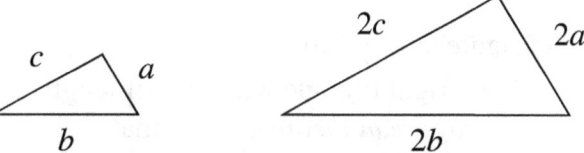

    · Equivalently, two triangles are *similar* if their corresponding angles are equal

    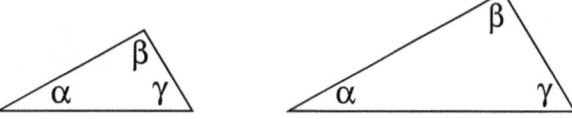

    · Equivalently, two triangles are *similar* if one is a magnified version of the other
    · A triangle with equal sides is an *equilateral triangle*

    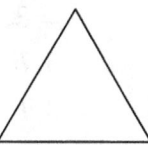

    · Equivalently, a triangle with all angles measuring 60° is an *equilateral triangle*
    · A triangle with *exactly* two equal sides is an *isosceles triangle*

    · Equivalently, a triangle with *exactly* two equal angles is an *isosceles triangle*
    · A triangle with unequal sides is a *scalene triangle*

    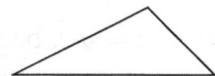

    · Equivalently, a triangle with unequal angles is a *scalene triangle*
    · An *acute triangle* has three acute angles

    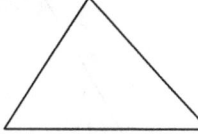

* Right triangles
    · A triangle with one angle measuring 90° is a *right triangle*
    · The side opposite the 90° angle in a right triangle is the *hypotenuse*
    · The sides adjacent to the 90° angle in a right triangle are the *legs*
    · The *hypotenuse* is always the longest side of a right triangle

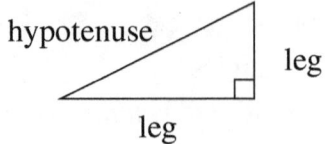

* Pythagorean theorem
    - For a right triangle with legs of length a and b and hypotenuse of length c, the *Pythagorean theorem* states that
    $$a^2 + b^2 = c^2$$

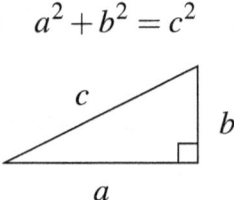

    - Three well-known integer values satisfying the Pythagorean theorem are
    $a = 3, b = 4, c = 5$, because $3^2 + 4^2 = 5^2$
    $a = 5, b = 12, c = 13$, because $5^2 + 12^2 = 13^2$
    $a = 8, b = 15, c = 17$, because $8^2 + 15^2 = 17^2$

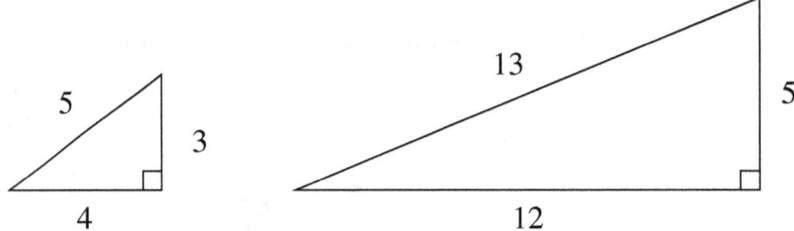

    - Two well-known non-integer values satisfying the Pythagorean theorem are
    $a = 1, b = \sqrt{3}, c = 2$, because $1^2 + \left(\sqrt{3}\right)^2 = 2^2$ (a 30°, 60°, 90° triangle)
    $a = 1, b = 1, c = \sqrt{2}$, because $1^2 + 1^2 = \left(\sqrt{2}\right)^2$ (a 45°, 45°, 90° triangle)

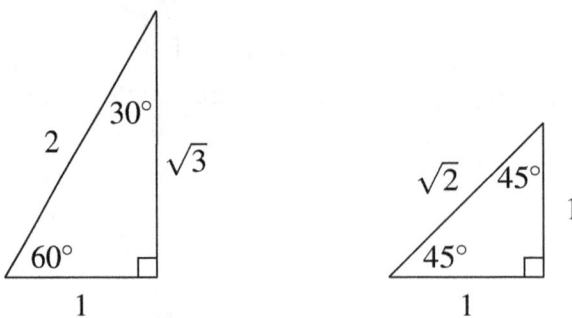

- Quadrilaterals
    * A *quadrilateral* is a four-sided polygon
    * Special quadrilaterals

Chapter 3. Geometry

· rectangle: a quadrilateral with four right angles

· square: a rectangle with equal side lengths

· parallelogram: a quadrilateral with parallel opposite sides

· rhombus: a parallelogram with four equal side lengths

· trapezoid: a quadrilateral with one pair of parallel sides

- Circles
    - A *circle* is a set of points that are an equal distance (called the *radius*) from a given point (called the *center*)

    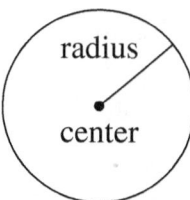

    - A circle can be thought of as the limit of a regular polygon as the number of sides goes to infinity
    - A circle is a two-dimensional object
    - A line segment whose endpoints are on a circle is a *chord*

    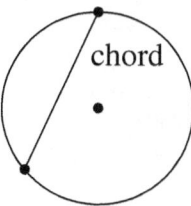

    - A chord that passes through the center of the circle is a *diameter*

    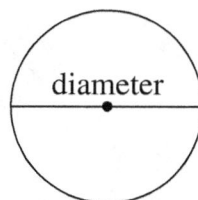

    - The length of a diameter is also known as the *diameter* of the circle

- The diameter is twice the radius
- An angle whose vertex is the center of a circle is known as a *central angle of a circle*

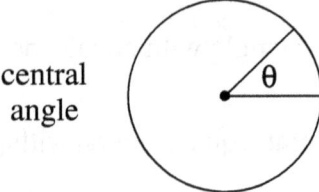

- All central angles divide a circle into a *major arc* and a *minor arc*

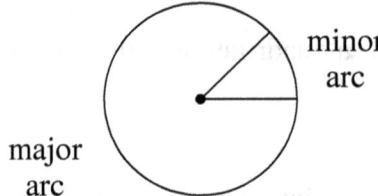

- A central angle of a circle that measures 180° divides the circle into two *semicircles*

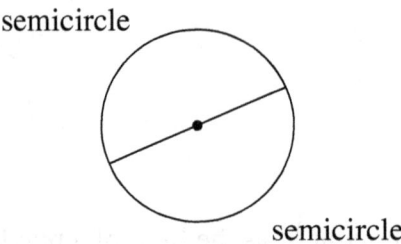

- An angle whose vertex is on the circle and whose sides are chords of the circle is known as an *inscribed angle*. The measure of an inscribed angle is one half of the associated central angle.

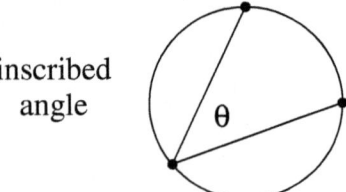

- A *circumscribed circle* passes through all vertices of a convex polygon

Chapter 3. Geometry

- *Concentric circles* have the same center but unequal radii

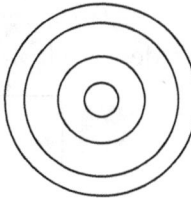

- Perimeter and circumference

  - The *perimeter* of a polygon is the sum of the lengths of the sides, for example, the perimeter of a square with sides of length s is $P = 4s$
  - The units of the perimeter of a polygon are the same as the units associated with the lengths of the sides
  - The *circumference* of a circle with radius r is

  $$C = 2\pi r$$

- Area

  There are formulas for the areas of several common geometric objects. The units of the area are the square of the units associated with the lengths of the sides.

  - *Area of a triangle*

  $$A = \frac{1}{2}bh$$

  where b and h are the lengths of the base and height of the triangle

  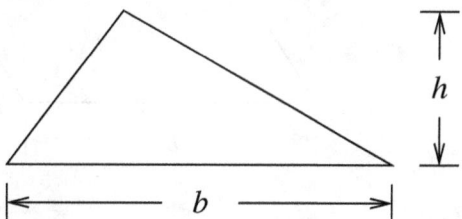

  - *Heron's formula*: the area of a triangle with sides of length a, b, and c is

  $$A = \sqrt{s(s-a)(s-b)(s-c)}$$

  where $s = (a+b+c)/2$ is the semi-perimeter

  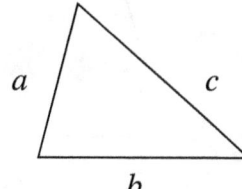

– *Area of a square*
$$A = s^2$$
where s is the length of a side

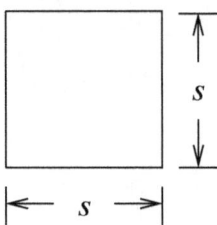

– *Area of a rectangle*
$$A = bh$$
where b and h are the lengths of the base and height of the rectangle

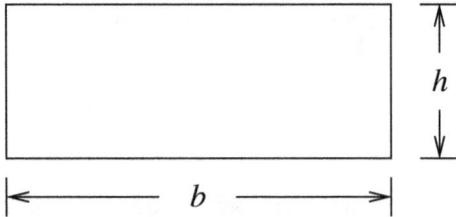

– *Area of a parallelogram*
$$A = bh$$
where b and h are the lengths of the base and height of the parallelogram

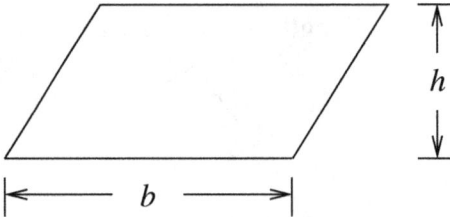

– *Area of a trapezoid*
$$A = \frac{a+b}{2} \cdot h$$
where a and b are the lengths of the two parallel sides of the trapezoid

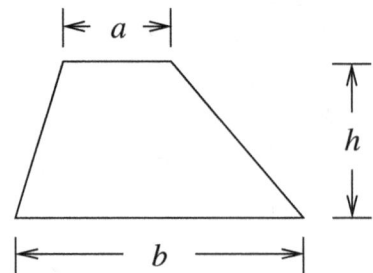

Chapter 3. Geometry

- *Area of a circle*

$$A = \pi r^2$$

where r is the radius of the circle

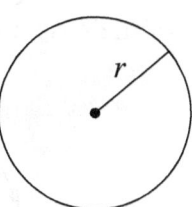

- The rectangle of fixed perimeter with the largest area is a square

- Solid geometry: three-dimensional objects

    - Surface area

        There are formulas for the surface area of several common geometric objects. The units of the surface area are the square of the units associated with the lengths of the sides.

        * *Surface area of a rectangular solid*

        $$A = 2(lw + lh + wh)$$

        where l, w, and h are the measures of the length, width, and height

        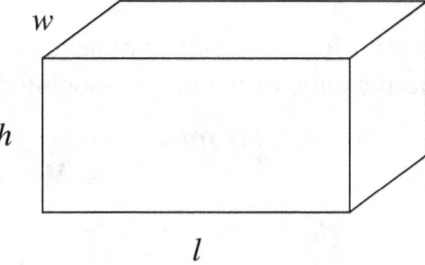

        * *Surface area of a (right circular) cylinder*

        $$A = 2\pi r^2 + 2\pi rh$$

        where r is the radius of the base and h is the height of the cylinder

        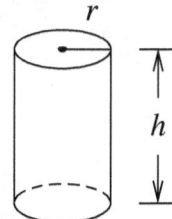

* *Surface area of a (right circular) cone*
$$A = \pi r^2 + \pi rs$$
where r is the radius of the base and s is the slant height

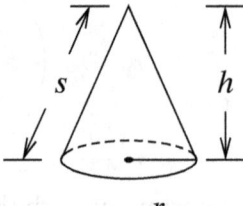

* *Surface area of a sphere*
$$A = 4\pi r^2$$
where r is the radius of the sphere

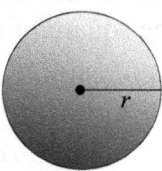

- Volume

  There are formulas for the volumes of several common geometric objects. The units of the volume are the cube of the units associated with the lengths of the sides.

  * *Volume of a rectangular solid*
  $$V = lwh$$

  * *Volume of a (right circular) cylinder*
  $$V = \pi r^2 h$$

  * *Volume of a (right circular) cone*
  $$V = \frac{1}{3}\pi r^2 h$$

  * *Volume of a sphere*
  $$V = \frac{4}{3}\pi r^3$$

Chapter 3. Geometry

Exercises

3.1 A fence is placed around a rectangular garden measuring 16 feet by 25 feet. How much fencing is necessary to enclose the garden?

3.2 Four distinct points A, B, C, and D lie on a line, ordered from left to right. If the length of $\overline{AC}$ is 10 and the length of $\overline{BD}$ is 7, what are the possible lengths of $\overline{BC}$?

3.3 A unit cube (each side of the cube has side length 1) is inscribed within a sphere such that each vertex of the cube lies on the sphere. What is the volume of the sphere?

3.1 A fence is placed around a rectangular garden measuring 16 feet by 25 feet. How much fencing is necessary to enclose the garden?

Since the perimeter of a rectangle is the sum of the lengths of the four sides, the amount of fencing required is
$$16 + 16 + 25 + 25 = 82$$
feet.

3.2 Four distinct points A, B, C, and D lie on a line, ordered from left to right. If the length of $\overline{AC}$ is 10 and the length of $\overline{BD}$ is 7, what are the possible lengths of $\overline{BC}$?

The length of $\overline{BC}$ could be as small as almost zero when points B and C are adjacent. The length of $\overline{BC}$ could be as large as almost 7 when points C and D are adjacent. So if x is the length of $\overline{BC}$, the values of x lie in the open interval
$$0 < x < 7$$
because the points A, B, C, and D are distinct.

3.3 A unit cube (each side of the cube has side length 1) is inscribed within a sphere such that each vertex of the cube lies on the sphere. What is the volume of the sphere?

The distance between opposite vertices of the cube is the diameter of the sphere d, as shown below. A first application of the Pythagorean theorem determines the length of a diagonal across one *side* of the cube to be $\sqrt{2}$ because
$$1^2 + 1^2 = \left(\sqrt{2}\right)^2.$$

A second application of the Pythagorean theorem with the hypotenuse of the triangle being the line segment that connects opposite vertices of the cube is
$$1^2 + \left(\sqrt{2}\right)^2 = d^2,$$
which gives $d = \sqrt{3}$ for the diameter of the sphere. The radius of the sphere is half of the diameter, that is, $r = \sqrt{3}/2$. Finally, the volume of the sphere is
$$V = \frac{4}{3}\pi r^3 = \frac{4}{3}\pi \left(\frac{\sqrt{3}}{2}\right)^3 = \frac{\pi}{2}\sqrt{3} \cong 2.72.$$

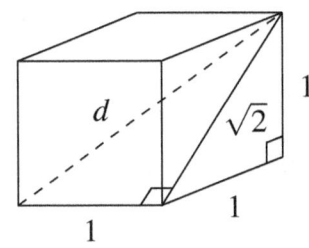

3.4 The radius of the larger circle below is twice the radius of the smaller circle. What is the ratio of the area of the larger circle to the area of the smaller circle?

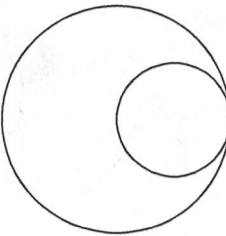

3.5 The diagram below consists of a circle and an equilateral triangle. One vertex of the equilateral triangle lies on the circle and the opposite side of the equilateral triangle is tangent to the circle. If the area of the circle is 1, what is the area of the triangle?

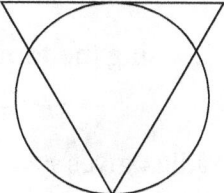

3.4 The radius of the larger circle below is twice the radius of the smaller circle. What is the ratio of the area of the larger circle to the area of the smaller circle?

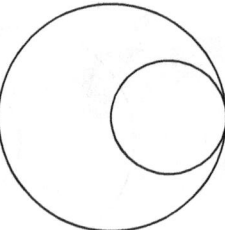

Without loss of generality, let r be the radius of the smaller circle. Then $2r$ is the radius of the larger circle. The area of the smaller circle is πr^2. The area of the larger circle is $\pi(2r)^2 = 4\pi r^2$. The ratio of the area of the larger circle to the area of the smaller circle is

$$\frac{4\pi r^2}{\pi r^2} = 4,$$

which can be expressed as the ratio 4:1.

3.5 The diagram below consists of a circle and an equilateral triangle. One vertex of the equilateral triangle lies on the circle and the opposite side of the equilateral triangle is tangent to the circle. If the area of the circle is 1, what is the area of the triangle?

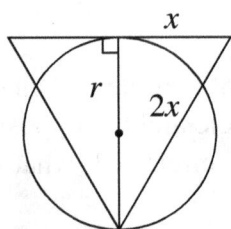

Let r be the radius of the circle. Using the formula for the area of a circle,

$$\pi r^2 = 1.$$

Solving this equation for the radius gives $r = 1/\sqrt{\pi}$. The associated diameter is twice the radius: $d = 2/\sqrt{\pi}$. Let $2x$ be the length of each side of the equilateral triangle. Adding a vertical diameter to the circle divides the equilateral triangle into two $30°, 60°, 90°$ triangles with legs of length x and $2/\sqrt{\pi}$, and hypotenuse of length $2x$. Applying the Pythagorean theorem to one of the $30°, 60°, 90°$ triangles gives

$$x^2 + (2/\sqrt{\pi})^2 = 4x^2.$$

Solving for x gives $x = 2/\sqrt{3\pi}$. Finally, the area of the equilateral triangle is one half of the product of the base and the height of the equilateral triangle:

$$A = \frac{1}{2} \cdot \frac{4}{\sqrt{3\pi}} \cdot \frac{2}{\sqrt{\pi}} = \frac{4}{\pi\sqrt{3}} \cong 0.7351.$$

Chapter 4

Analytic Geometry

Analytic geometry uses a coordinate system in the study of geometry. The tools of algebra can then be applied to problems in geometry, effectively combining the previous two chapters. The rectangular coordinate system, which specifies a horizontal position and a vertical position relative to an origin, is the most commonly used system for defining points and curves.

- The Cartesian plane (a.k.a., the rectangular coordinate system)
    - The system is formed by intersecting a horizontal number line and a vertical number line at their origins
    - The horizontal number line is known as the x-axis
    - The vertical number line is known as the y-axis
    - The point of intersection of the two number lines is known as the *origin*
    - The x-axis is often labeled x
    - The y-axis is often labeled y
    - Quadrants
        * The x-axis and the y-axis divide the Cartesian plane into four *quadrants*, which are traditionally designated by Roman numerals
        * The region associated with $x > 0$ and $y > 0$ is known as quadrant I (or the first quadrant)
        * Quadrants II, III, and IV follow counterclockwise from quadrant I

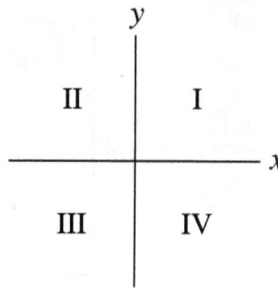

Chapter 4. Analytic Geometry

- Points
    - A *point* is denoted by the ordered pair (x, y)

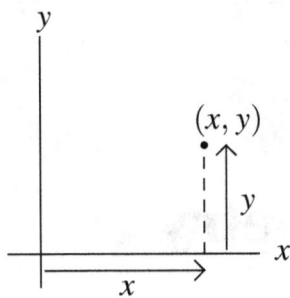

    - The point $(0, 0)$, where the two number lines intersect, is the *origin*
    - The *abscissa* is the *x*-coordinate, which is the directed horizontal distance from the origin
    - The *ordinate* is the *y*-coordinate, which is the directed vertical distance from the origin

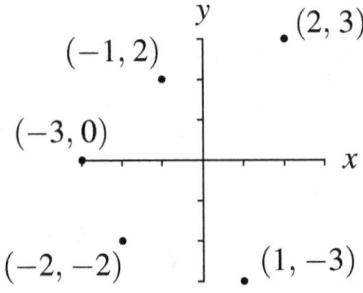

- Midpoint and distance formulas
    - *Midpoint formula:* the midpoint of the line segment from the point (x_1, y_1) to the point (x_2, y_2) is
    $$\left(\frac{x_1 + x_2}{2}, \frac{y_1 + y_2}{2}\right)$$

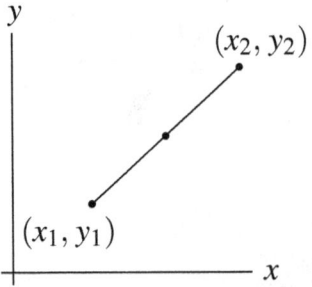

Chapter 4. Analytic Geometry

- *Distance formula:* using the Pythagorean theorem, the distance between the points (x_1, y_1) and (x_2, y_2) is

$$d = \sqrt{(x_2 - x_1)^2 + (y_2 - y_1)^2}$$

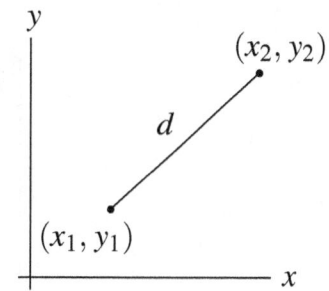

- Graphs and symmetry

  - The *graph* of an equation in the variables x and y is the set of all points (x, y) in the Cartesian plane satisfying the equation

  - Techniques for graphing an equation in the two variables x and y

    * Plotting points
      · solve for one variable (preferably y)
      · make a table of selected solution points
      · plot the solution points in the Cartesian plane
      · connect the points with a smooth curve
    * Graphing calculator
    * Computer program (for example, Maple, Mathematica, R)

  - The graph of an equation is *symmetric about the y-axis* if replacing x with $-x$ in the equation results in the original equation, that is, if (x, y) is a point on the graph of the equation, then $(-x, y)$ is also on the graph. Geometrically, every point on the graph has a mirror image on the opposite side of the y-axis. Three examples are shown below.

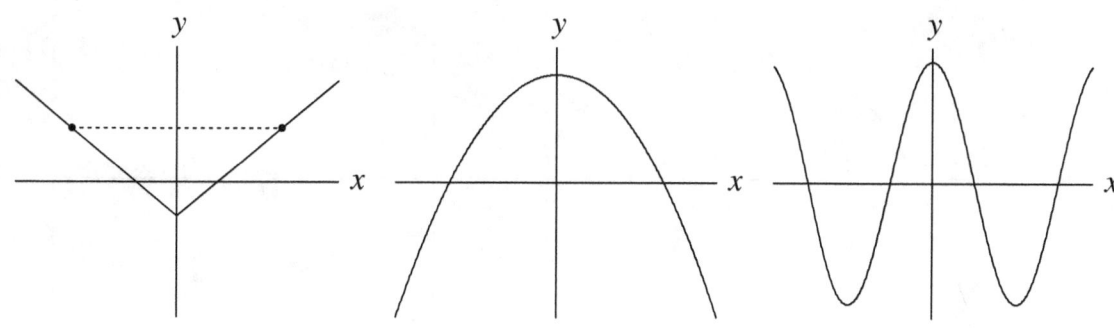

- The graph of an equation is *symmetric about the x-axis* if replacing y with $-y$ in the equation results in the original equation, that is, if (x, y) is a point on the graph of the equation, then $(x, -y)$ is also on the graph. Geometrically, every point on the graph has a mirror image on the opposite side of the x-axis. Three examples are shown below.

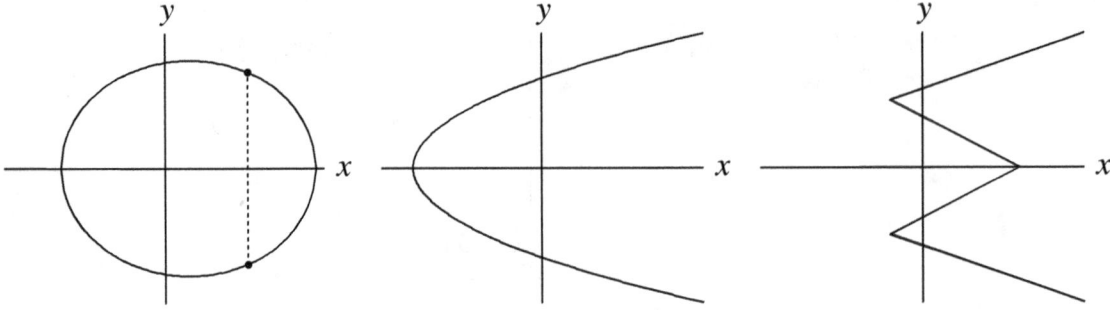

- The graph of an equation is *symmetric about the origin* if replacing x with $-x$ and y with $-y$ in the equation results in the original equation, that is, if (x, y) is a point on the graph of the equation, then $(-x, -y)$ is also on the graph. Geometrically, every point on the graph has a mirror image on the opposite side of the origin. Three examples are shown below.

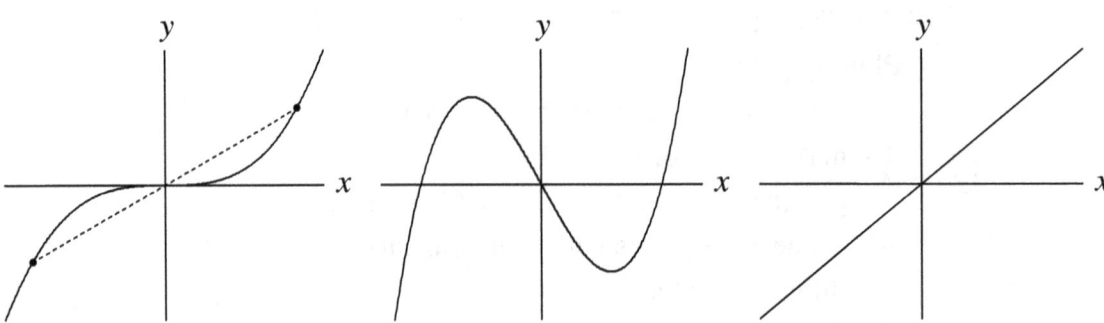

- Intercepts

    - A *y-intercept* in the graph of an equation in x and y is a point where the graph touches or crosses the y-axis. The y-intercepts of a graph are found by setting x equal to zero and solving the resulting equation for y. Graphs with one, two, and four y-intercept(s) are shown below.

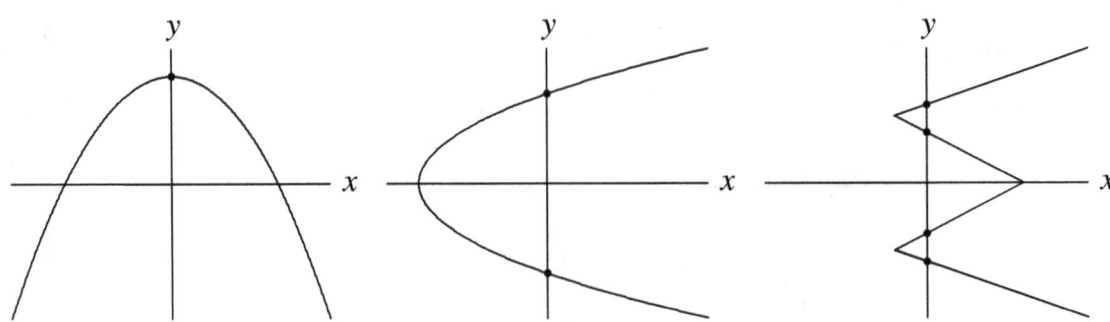

- An *x-intercept* in the graph of an equation in x and y is a point where the graph touches or crosses the x-axis. The x-intercepts of a graph are found by setting y equal to zero and solving the resulting equation for x. Graphs with three, two, and one x-intercept(s) are shown below.

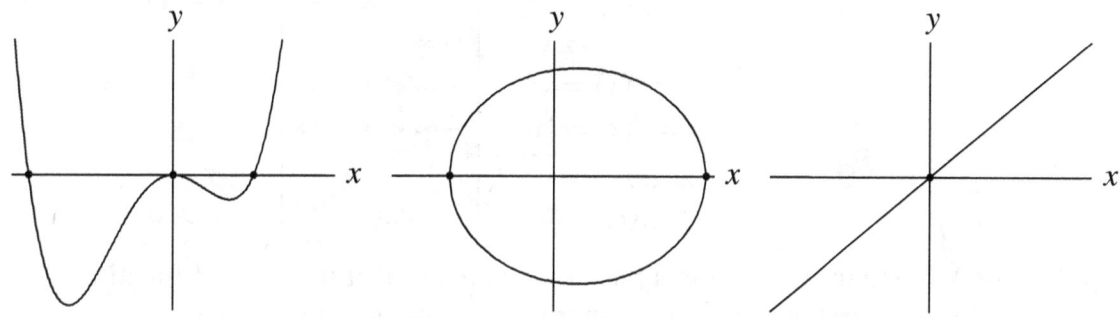

- Functions

  - A *function* $y = f(x)$ pairs each element in the *domain* (the set of values allowed for x) with exactly one element in the *range* (the set of values allowed for y)
  - Functional notation
    * f is the name of the function
    * x is the *independent* variable
    * y is the *dependent* variable
    * $f(x)$ is the value of the function associated with x, read as "f of x"
    * other letters can be substituted for x, y, and f
  - *Vertical line test*: A graph is associated with a *function* if a vertical line drawn at any x position intersects the graph at most once. If one or more vertical lines can be drawn that intersect the graph two or more times, then the graph does not correspond to a function.

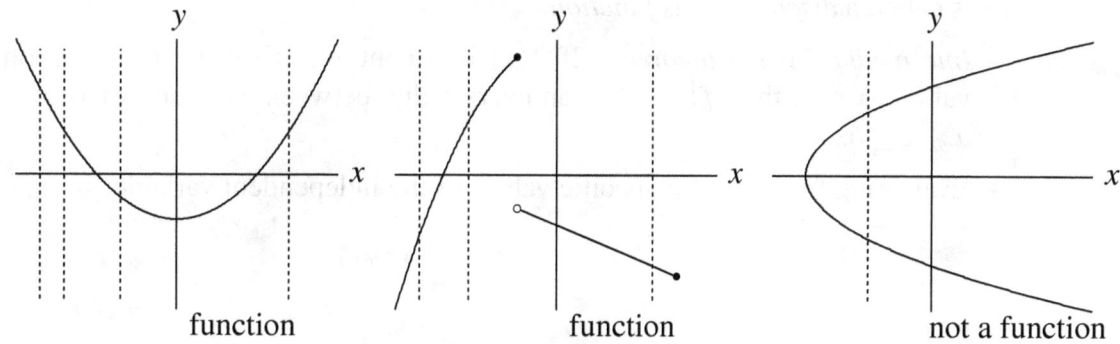

  - Examples of relationships between x and y that are not functions:
    * $x = y^2$
    * $x = |y|$
    * $x^2 + y^2 = 1$

- Examples of functions:

function name	domain	range
$y = f(x) = 3x - 5$	$-\infty < x < \infty$	$-\infty < y < \infty$
$y = f(x) = 17$	$-\infty < x < \infty$	$y = 17$
$y = f(x) = x^2$	$-\infty < x < \infty$	$y \geq 0$
$y = f(x) = x^3$	$-\infty < x < \infty$	$-\infty < y < \infty$
$y = f(x) = |x|$	$-\infty < x < \infty$	$y \geq 0$
$y = f(x) = \sqrt{x-4}$	$x \geq 4$	$y \geq 0$
$a = g(r) = \pi r^2$	$r > 0$	$a > 0$

- A function $y = f(x)$ is a *piecewise function* if it has several functional forms defined over several intervals that comprise the domain, for example,

$$f(x) = \begin{cases} -x - 3 & -2 \leq x < -1 \\ x^2 & -1 \leq x \leq 1/2 \\ 5/4 - 2x & x > 1/2 \end{cases}$$

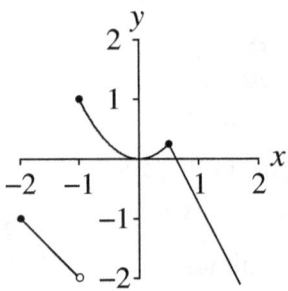

- A function is *continuous* if its graph can be drawn without lifting a pencil from the paper; in other words, it has no breaks or holes. A function that is that is not continuous is called a *discontinuous function*.

- *Intermediate value theorem.* If $f(x)$ is a continuous function defined on the interval $a \leq x \leq b$, then $f(x)$ takes on every value between $f(a)$ and $f(b)$ on the interval $a \leq x \leq b$.

- Evaluating functions for specific values of the independent variable: if

$$f(x) = x^2 + 1$$

then
  * $f(7) = 50$
  * $f(-2) = 5$
  * $f(0) = 1$
  * $f(c-3) = (c-3)^2 + 1$

Chapter 4. Analytic Geometry

- A *zero* of a function $y = f(x)$ is a value of x such that $f(x) = 0$. These values correspond to the x-intercepts of the graph of $y = f(x)$.
- Increasing and decreasing functions over intervals
    * A function $y = f(x)$ is *increasing* on an interval if $f(x_1) < f(x_2)$ for every $x_1 < x_2$ on the interval
    * A function $y = f(x)$ is *decreasing* on an interval if $f(x_1) > f(x_2)$ for every $x_1 < x_2$ on the interval
    * A function $y = f(x)$ is *constant* on an interval if $f(x_1) = f(x_2)$ for every $x_1 < x_2$ on the interval
    * A function $y = f(x)$ is *nondecreasing* on an interval if $f(x_1) \leq f(x_2)$ for every $x_1 < x_2$ on the interval
    * A function $y = f(x)$ is *nonincreasing* on an interval if $f(x_1) \geq f(x_2)$ for every $x_1 < x_2$ on the interval
- Increasing and decreasing functions
    * A function $y = f(x)$ is an *increasing function* if $f(x_1) < f(x_2)$ for every $x_1 < x_2$ in its domain
    * A function $y = f(x)$ is an *decreasing function* if $f(x_1) > f(x_2)$ for every $x_1 < x_2$ in its domain
    * A function $y = f(x)$ is an *constant function* if $f(x_1) = f(x_2)$ for every $x_1 < x_2$ in its domain
    * A function $y = f(x)$ is an *nondecreasing function* if $f(x_1) \leq f(x_2)$ for every $x_1 < x_2$ in its domain
    * A function $y = f(x)$ is an *nonincreasing function* if $f(x_1) \geq f(x_2)$ for every $x_1 < x_2$ in its domain
- Common functions

$y = f(x) = a$	constant (for some real-valued constant a)
$y = f(x) = x$	linear (slope 1, y-intercept 0) identity function
$y = f(x) = \|x\|$	absolute value
$y = f(x) = x^2$	quadratic
$y = f(x) = x^3$	cubic
$y = f(x) = x^{-1} = 1/x$	reciprocal ($x \neq 0$)
$y = f(x) = x^{1/2} = \sqrt{x}$	square root ($x \geq 0$)
$y = f(x) = x^{1/3} = \sqrt[3]{x}$	cube root
$y = f(x) = a^x$	power (for some real-valued positive constant $a \neq 1$)
$y = f(x) = \log x$	logarithmic ($x > 0$)
$y = f(x) = \lfloor x \rfloor$	floor (greatest integer less than or equal to x)
$y = f(x) = \lceil x \rceil$	ceiling (least integer greater than or equal to x)

The generic shapes of these twelve functions are shown in the 4 × 3 display of functions (reading row-wise) with positive values for *a*.

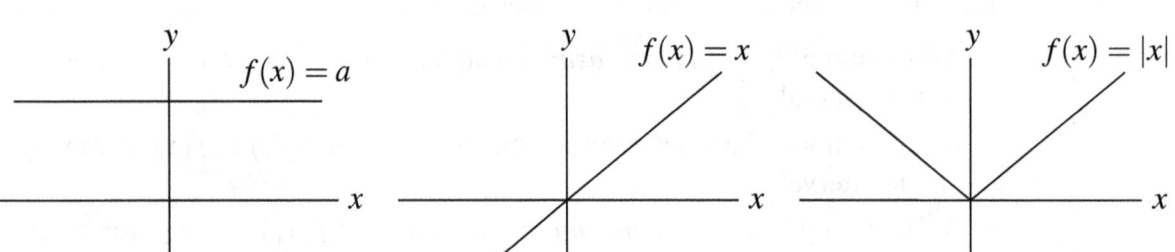
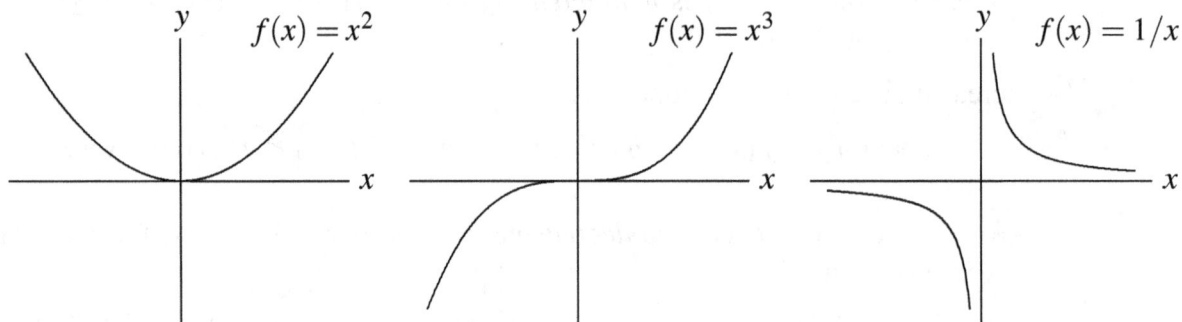
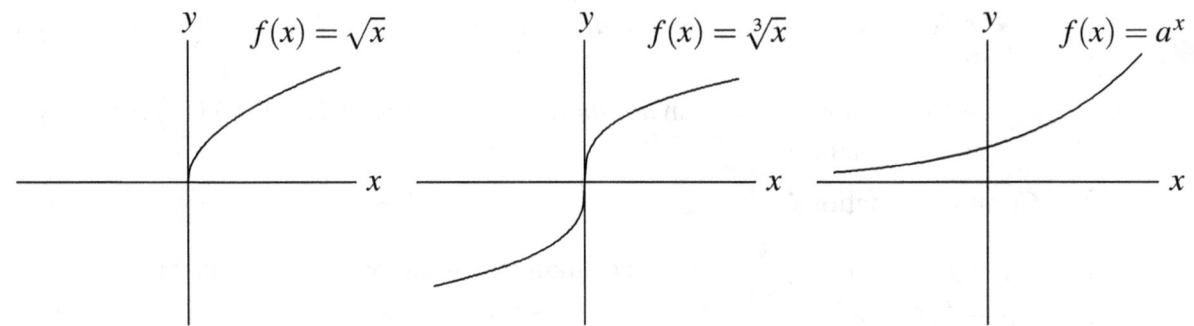
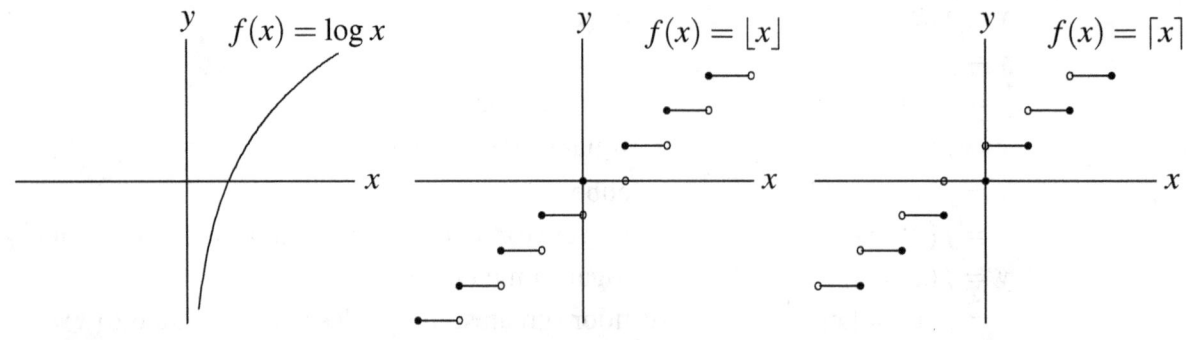

Chapter 4. Analytic Geometry

- The graph of a function has an *asymptote* if a line exists such that the distance between the graph of the function and the line approaches zero as both tend to infinity. For example, the x-axis is a horizontal asymptote and the y-axis is a vertical asymptote for the reciprocal function.

- Transforming functions: the effect on the graph of $f(x)$ of shifting, reflecting, scaling, etc., treated generically below with the constant 2

$f(x)+2$	*vertical shift* of the graph up two units
$f(x)-2$	*vertical shift* of the graph down two units
$f(x+2)$	*horizontal shift* of the graph left two units
$f(x-2)$	*horizontal shift* of the graph right two units
$-f(x)$	*reflect* the graph about x-axis
$f(-x)$	*reflect* the graph about y-axis
$2f(x)$	*vertical stretch* of the graph by a factor of two
$\frac{1}{2}f(x)$	*vertical shrink* of the graph by a factor of two
$f(2x)$	*horizontal shrink* of the graph by a factor of two
$f\left(\frac{1}{2}x\right)$	*horizontal stretch* of the graph by a factor of two

The first six transformations are known as *rigid transformations* because the shape of the graph is unchanged. The first four transformations are *shifts* or *translations* and the next two transformations are *reflections*. These transformations can be applied in a serial fashion, for example, if $f(x) = |x|$, then the graph of

$$g(x) = |x+1| - 3 = f(x+1) - 3$$

is a horizontal shift of the parent function $f(x)$ (dashed curve) left one unit (dotted curve), then a vertical shift downward three units (solid curve).

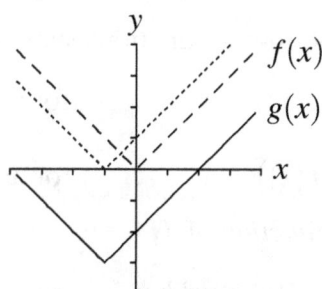

- Classifying functions

    - If $f(x) = f(-x)$ for every x in the domain of f, then f is an *even function* and its graph is symmetric about the y-axis
    - If $f(-x) = -f(x)$ for every x in the domain of f, then f is an *odd function* and its graph is symmetric about the origin

- Combining functions
    - Arithmetic operations (The domain of these functions is the intersection of the domain of f and the domain of g. In addition, $g(x) \neq 0$ for division.)
        * Adding functions: $(f+g)(x) = f(x) + g(x)$
        * Subtracting functions: $(f-g)(x) = f(x) - g(x)$
        * Multiplying functions: $(f \cdot g)(x) = f(x) \cdot g(x)$
        * Dividing functions: $(f/g)(x) = f(x)/g(x)$
    - Composition (The domain of $f(g(x))$ is the set of all x in the domain of g such that $g(x)$ is in the domain of f.)
        $$(f \circ g)(x) = f(g(x))$$

- One-to-one functions
    - A function $f(x)$ is a *one-to-one function* (also written 1–1) if for every pair of distinct values x_1 and x_2 in the domain of $f(x)$, $f(x_1) \neq f(x_2)$
    - Equivalently, a function $y = f(x)$ is a one-to-one function if each value of the dependent variable comes from exactly one value of the independent variable
    - *Horizontal line test*: A function $f(x)$ is a one-to-one function if its graph is never intersected by any horizontal line more than once
    - Examples:
        * $x = |y|$ is not a function
        * $y = |x|$ is a function, but is not one-to-one
        * $y = x$ is a function and is one-to-one

- Inverse functions
    - If f and g are two one-to-one functions such that
        $$f(g(x)) = x \qquad \text{for any } x \text{ in the domain of } g$$
      and
        $$g(f(x)) = x \qquad \text{for any } x \text{ in the domain of } f$$
      then g is the *inverse function* of f, for example,
        $$f(x) = x^3 \text{ has inverse function } g(x) = x^{1/3}$$
    - The inverse function of f is often denoted by f^{-1}. Warning: $f^{-1}(x)$ is not $1/f(x)$.
    - An inverse function essentially undoes the operation performed by the original function, for example,
        $$f(x) = x^3 \qquad \Rightarrow \qquad f^{-1}(f(x)) = f^{-1}(x^3) = \sqrt[3]{x^3} = x$$
    - The domain of $f^{-1}(x)$ is the range of $f(x)$

- The domain of $f(x)$ is the range of $f^{-1}(x)$
- To find the inverse of a one-to-one function $y = f(x)$, exchange the roles of x and y in the equation, that is, $x = f(y)$, and solve for y, yielding $y = f^{-1}(x)$
- Two well-known functions satisfying the relationship $f(x) = f^{-1}(x)$ are the reciprocal function $f(x) = 1/x$ and the identity function $f(x) = x$
- Equivalent statements
    * $f(x)$ has an inverse
    * $f(x)$ is a one-to-one function
    * $f(x)$ passes the vertical and horizontal line tests
- The graph of $f^{-1}(x)$ is the reflection of $f(x)$ across the line $y = x$

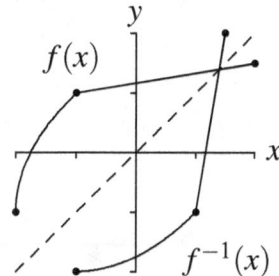

- Linear functions (lines)

    - Slope

        * The *slope m* for the line passing through the points (x_1, y_1) and (x_2, y_2) is

        $$m = \frac{y_2 - y_1}{x_2 - x_1}$$

        * The slope is also known as "rise over run" or "change in y over change in x." This "rate of change" is measured in vertical units per horizontal units.

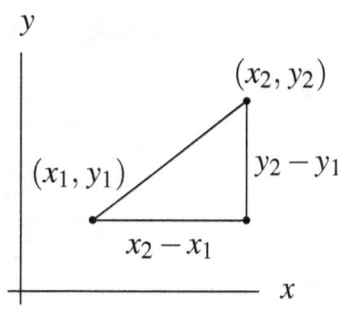

        * When $m > 0$, the line is rising (in other words, the line runs uphill)
        * When $m < 0$, the line is falling (in other words, the line runs downhill)

* When $m = 0$, the line is horizontal
            * The slope m is undefined for a vertical line
        - Formulas for lines
            * The *general formula* for a line is
            $$ax + by = c$$
            where a, b, and c are real constants (b is *not* the y-intercept). The coefficients a and b can't both be 0.
            * The *slope–intercept formula* for a line with slope m and y-intercept b is
            $$y = mx + b$$
            This is the best formula for plotting a line. The y-intercept is identified first, then the "rise over run" interpretation of the slope is used to plot the line.

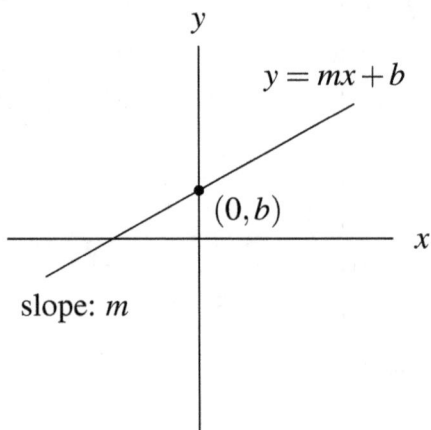

            * The *point–slope formula for a line* with slope m passing through the point (x_1, y_1) is
            $$y - y_1 = m(x - x_1)$$
            This formula for a line can be derived from the definition of the slope by letting $y = y_2$ and $x = x_2$ and rearranging terms.
        - The line $y = b$ is a horizontal line with slope $m = 0$

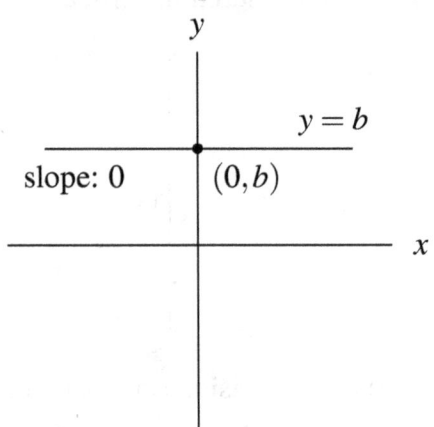

Chapter 4. Analytic Geometry

- The line $x = a$ is a *vertical line* with undefined slope

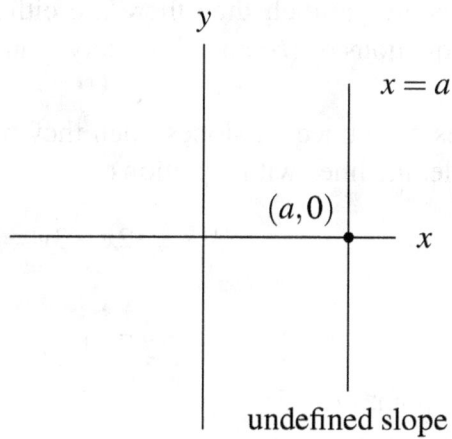

- Two nonvertical *parallel lines* have the same slope

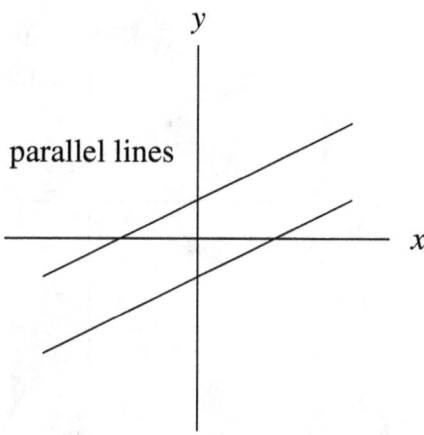

- Two nonvertical *perpendicular lines* with nonzero slopes have opposite reciprocal slopes

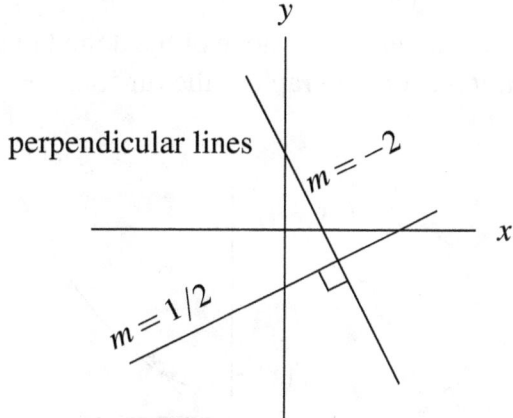

- Finding the intersection point of two lines
  * If two lines are parallel, then there are either (*a*) no solutions if the lines have different equations or (*b*) have infinitely many solutions if the lines have identical equations
  * If two lines have unequal slopes, then they meet at a single point of intersection, for example, the lines with equations

$$2x - 3y = 8$$

and

$$7x + 2y = 3$$

  meet at the point $(1, -2)$
- Through any two distinct points, there passes one and only one line

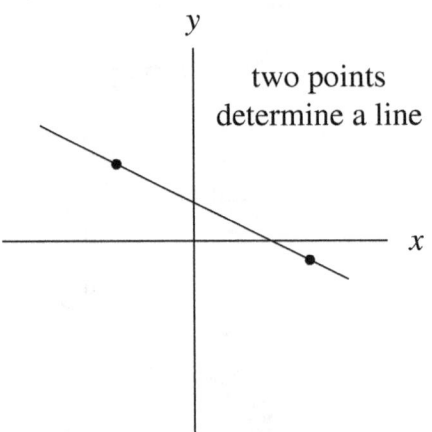

- The *difference quotient* is the slope of the line passing through the points $(x, f(x))$ and $(x+h, f(x+h))$:

$$\frac{f(x+h) - f(x)}{h}$$

The difference quotient is the slope of the dotted line in the figure below, which is often called a *secant line* of the graph of the function.

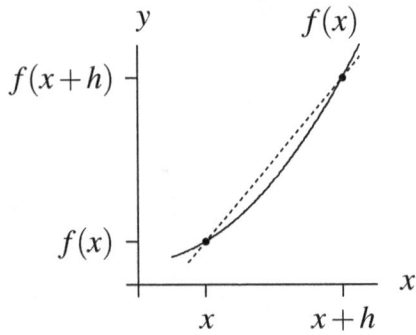

Chapter 4. Analytic Geometry

- Quadratic functions
    - The graph of a quadratic function is called a *parabola*
    - All parabolas are symmetric about a vertical line known as the *axis of symmetry*
    - The *standard form* for a quadratic function is

    $$y = f(x) = a(x-h)^2 + k$$

    where the point (h, k) is the *vertex* of the parabola. The vertical line $x = h$ is the axis of symmetry. This is the best form for drawing a graph of the quadratic function.
    - The *general form* for a quadratic function is

    $$y = f(x) = ax^2 + bx + c$$

    for real numbers $a \neq 0$, b, and c. The point

    $$\left(\frac{-b}{2a}, f\left(\frac{-b}{2a}\right)\right)$$

    is the vertex of the parabola.
    - Switching between the standard and general forms
        * To convert the standard form to the general form, multiply out the squared term and simplify
        * To convert the general form to the standard form, complete the square, for example,

    $$\begin{aligned} f(x) &= x^2 + 6x + 7 \\ &= \left(x^2 + 6x + 9\right) + 7 - 9 \\ &= (x+3)^2 - 2 \end{aligned}$$

    - If $a > 0$ the parabola opens upward (in other words, the parabola is "concave up")

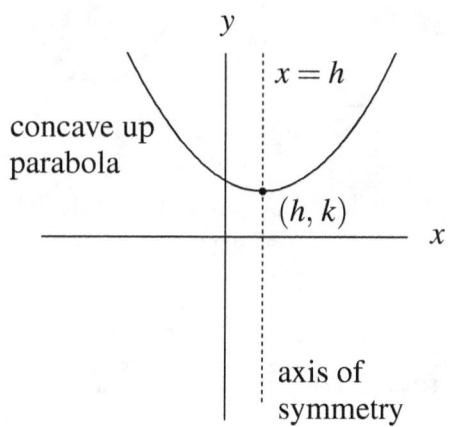

- If $a < 0$ the parabola opens downward (in other words, the parabola is "concave down")

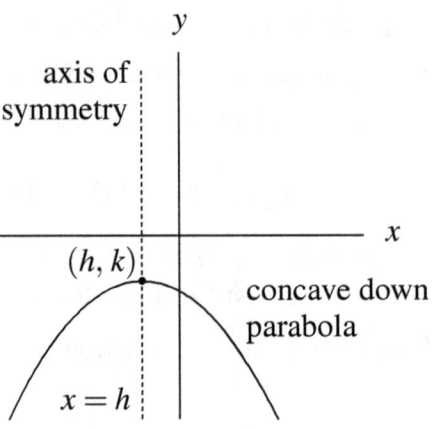

- Intercepts
    * Parabolas have a single y-intercept $y = f(0)$
    * The graphs of quadratic functions have zero, one, or two x-intercepts which can be found by solving the associated general from of the quadratic equation
    $$ax^2 + bx + c = 0$$
- Through any three points not on the same line, there passes one and only one parabola

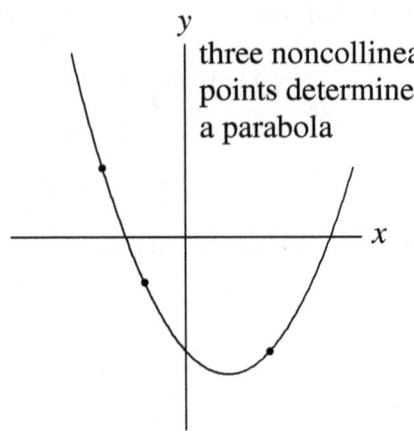

- Applications: quadratic functions are used to design reflectors and model the motion of an object under the force of gravity

- Polynomial functions
    - The standard form for a degree n *polynomial function* in x is
    $$y = f(x) = a_n x^n + a_{n-1} x^{n-1} + \cdots + a_2 x^2 + a_1 x + a_0$$

Chapter 4. Analytic Geometry

where the coefficients $a_n, a_{n-1}, \ldots, a_2, a_1, a_0$ are real constants, $a_n \neq 0$ is the *leading coefficient*, and a_0 is the *constant term*

- A degree n polynomial function in x with a single term

$$y = f(x) = a_n x^n$$

is an even function when n is even, and it is an odd function when n is odd

- The graph of a polynomial function is a continuous curve that makes no sharp turns
- The y-intercept of the graph of a polynomial function is

$$y = f(0) = a_0$$

that is, the graph of the polynomial function passes through $(0, a_0)$

- A real-valued *root* or *solution* to the polynomial equation

$$y = f(x) = a_n x^n + a_{n-1} x^{n-1} + \cdots + a_2 x^2 + a_1 x + a_0 = 0$$

corresponds to an x-intercept of the associated graph

- A degree n polynomial has, at most, n real-valued zeros
- If a polynomial can be factored, then the zeros can easily be determined by setting each factor equal to zero, for example, the degree seven ($n = 7$) polynomial

$$y = f(x) = x(x+1)^2 (x-2)(x-4)^3$$

has seven factors and seven real zeros: $x = 0, -1, -1, 2, 4, 4, 4$

- The *multiplicity* of a zero of a polynomial is the number of times that the polynomial has that particular zero. For the previous polynomial, the zero $x = 4$ has multiplicity 3.
- Interpreting factors
  * A factor of a polynomial of the form $(x-a)^k$ corresponds to the graph of the polynomial function *crossing* the x-axis at $x = a$ when k is *odd*
  * A factor of a polynomial of the form $(x-a)^k$ corresponds to the graph of the polynomial function *touching* the x-axis at $x = a$ when k is *even*
- The limiting behavior of a polynomial function as $x \to \infty$ or as $x \to -\infty$ is the same as the behavior of the polynomial function with identical leading coefficient:

$$y = f(x) = a_n x^n$$

So the sign of the leading coefficient and n determine the behavior of a polynomial function at its extremes.

- Polynomials can be divided in an analogous fashion to integers via *long division* (which works generally) or *synthetic division* (which works for divisors of the form $x - a$, where $a \neq 0$)

- Possible rational zeros. Every rational zero of a degree n polynomial with integer coefficients $a_n, a_{n-1}, \ldots, a_1, a_0$ is the ratio of a factor of a_0 to a factor of a_n, for example, $f(x) = 5x^3 - 3x^2 + 25x - 15$ has possible rational zeros: $\pm 15, \pm 5, \pm 3, \pm 1, \pm \frac{3}{5}, \pm \frac{1}{5}$
- Descartes' rule of signs. For a polynomial written in standard form with real coefficients, the number of distinct positive zeros of the polynomial equals the number of sign differences between consecutive nonzero coefficients, or is less than this count by a multiple of 2. Multiple roots of the same value are counted by their multiplicity. The polynomial $x^3 - x^2 - x - 1$, for example, has just one sign change, so it must have exactly one distinct positive root.

- Rational functions

  - A rational function is the ratio of two polynomial functions
  - Letting $N(x)$ be the polynomial in the numerator and $D(x)$ be the polynomial in the denominator, the standard form for a rational function $f(x)$ is

  $$f(x) = \frac{N(x)}{D(x)} = \frac{a_n x^n + a_{n-1} x^{n-1} + \cdots + a_2 x^2 + a_1 x + a_0}{b_m x^m + b_{m-1} x^{m-1} + \cdots + b_2 x^2 + b_1 x + b_0}$$

  - If $N(x)$ and $D(x)$ have no zeros in common, the graph of a rational function $y = f(x)$ has vertical asymptotes at all x values associated with $D(x) = 0$
  - Horizontal and slant asymptotes on the graph of a rational function $y = f(x)$ can be determined by n and m, the degrees of the two polynomials
    * There is a horizontal asymptote at $y = 0$ when $n < m$
    * There is a horizontal asymptote at $y = a_n/b_m$ when $n = m$
    * There is a slant asymptote with slope a_n/b_m when $n = m + 1$ (the equation of the slant asymptote can be determined by dividing $N(x)$ by $D(x)$ and ignoring the fractional part of the result)
    * There are neither horizontal asymptotes nor slant asymptotes when $n > m + 1$

- Exponential functions

  - An *exponential function* in x has the form

  $$y = f(x) = a^x$$

  where a is the *base* satisfying $a > 0$ and $a \neq 1$
  - Exponential functions are one-to-one functions
  - Exponential functions are positive for all values of x
  - Exponential functions are *decreasing* when $0 < a < 1$
  - Exponential functions are *increasing* when $a > 1$

Chapter 4. Analytic Geometry

- The graph of an exponential function always passes through the point (0, 1) because $y = f(0) = a^0 = 1$
- The graph of an exponential function has one of two characteristic shapes

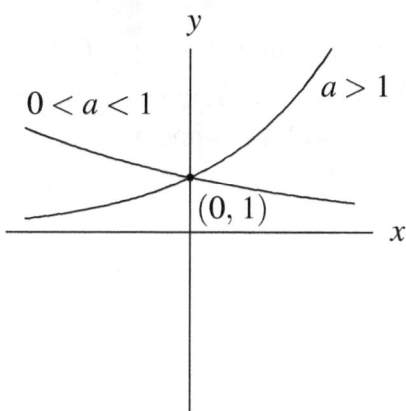

- The exponential function
$$y = f(x) = e^x$$
has the "natural" base $e = 2.7182818284590\ldots$ (Euler's number), where
$$e = \lim_{n \to \infty} \left(1 + \frac{1}{n}\right)^n$$

- The line that is tangent to the graph of $y = e^x$ at (0, 1) has slope 1
- Exponential functions are used to model decay ($0 < a < 1$) and growth ($a > 1$), for example,
    * population growth
    * radioactive decay
    * compound interest, for example, continuously compounded interest. The dollar amount A after t years of continuous compounding interest on an initial principal amount P at an annual interest rate r is
$$A = Pe^{rt}$$

- Logarithmic functions

    - A *logarithmic function* in x has the form
$$y = f(x) = \log_b x$$
    where the positive real number b is the *base*

- Logarithmic functions are one-to-one functions
- Logarithmic functions are *increasing* functions
- The graph of a logarithmic function always passes through the point $(1,0)$ because $y = f(1) = \log_b 1 = 0$
- The inverse of the logarithmic function $f(x) = \log_b x$ is $f^{-1}(x) = b^x$
- The graph of a logarithmic function is the reflection of the graph of the associated exponential function with the same base across the line $y = x$

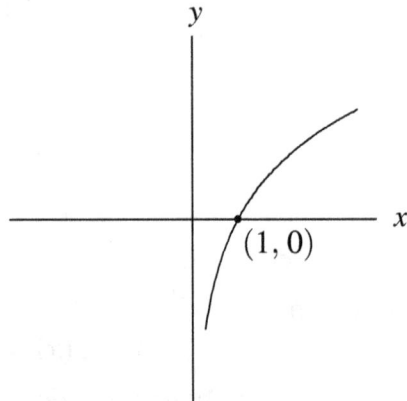

- The two most common bases are base 10 (common logarithms) and base e (natural logarithms)
    * The *common logarithmic function* corresponds to a logarithm base 10. The base is often dropped from the notation, that is, the function $y = f(x) = \log x$ is the same as the function $y = f(x) = \log_{10} x$
    * The *natural logarithmic function* corresponds to a logarithm base e and is written as $y = f(x) = \ln x$

- **Conic sections**

    Conic sections are curves obtained by intersecting a double-napped cone and a plane. Alternatively, conic sections can be obtained by considering distances, as indicated below.

    - Circle
        * A circle is the set of points that are a fixed distance r (the radius) from a center point (h, k), that is, using the distance formula $\sqrt{(x-h)^2 + (y-k)^2} = r$
        * The *standard form* for a circle of radius r centered at (h, k) is

        $$(x-h)^2 + (y-k)^2 = r^2$$

        The standard form for a circle is the best arrangement for plotting the circle on a Cartesian coordinate system.

Chapter 4. Analytic Geometry

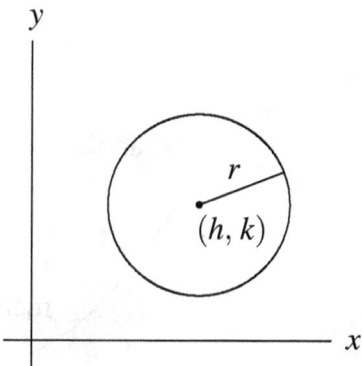

* The *general form* for a circle is

$$x^2 + y^2 + ax + by + c = 0$$

* A circle is *not* the graph of a function because there are two y-values associated with all but two of the vertical lines that intersect the circle
* A circle with radius 1 and center $(0, 0)$ with equation

$$x^2 + y^2 = 1$$

is called a *unit circle*
* Through any three points not on the same line, there passes one and only one circle

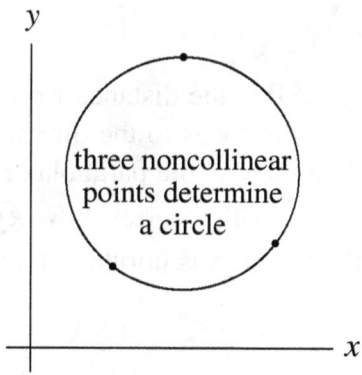

– Parabola

* Parabolas were considered earlier as the graphs associated with quadratic functions. They can also be considered more generally as conic sections.
* A parabola is the set of points that are equally distant from a line (known as the *directrix*) and a point (known as the *focus*) that is not on the line. The *vertex* of the parabola is the midpoint between the focus and the directrix. The line passing through the focus and the vertex is the *axis of symmetry*.

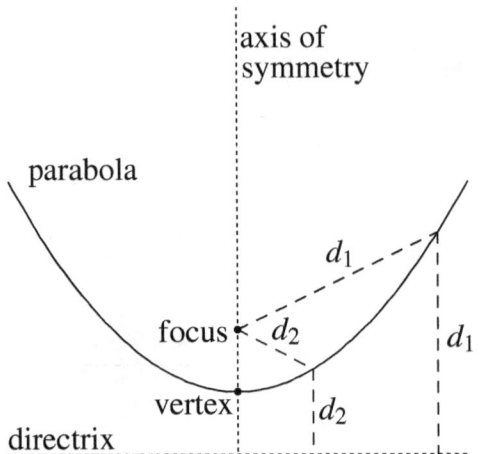

* When the directrix is horizontal, the parabola corresponds to a quadratic function
* When the directrix is vertical, the parabola does not correspond to a function
* Case I: Parabola with a vertex at the origin
  · When the directrix is horizontal, the parabola is described by

  $$x^2 = 4py$$

  where $|p| \neq 0$ is the distance from the focus to the vertex, as well as the distance from the vertex to the directrix. When $p > 0$, the parabola is concave up. When $p < 0$, the parabola is concave down.

  · When the directrix is vertical, the parabola is described by

  $$y^2 = 4px$$

  where $|p| \neq 0$ is the distance from the focus to the vertex, as well as the distance from the focus to the directrix. When $p > 0$, the parabola opens to the right. When $p < 0$, the parabola opens to the left.

* Case II: Parabola with a vertex at (h, k)
  · When the directrix is horizontal, the parabola is described by

  $$(x-h)^2 = 4p(y-k)$$

  where $|p| \neq 0$ is the distance from the focus to the vertex, as well as the distance from the focus to the directrix. When $p > 0$, the parabola is concave up. When $p < 0$, the parabola is concave down.

  · When the directrix is vertical, the parabola is described by

  $$(y-k)^2 = 4p(x-h)$$

  where $|p| \neq 0$ is the distance from the focus to the vertex, as well as the distance from the focus to the directrix. When $p > 0$, the parabola opens to the right. When $p < 0$, the parabola opens to the left.

Chapter 4. Analytic Geometry

* *Reflective property* of a parabola. All lines incoming to a parabola that are parallel to the axis of symmetry reflect off the parabola and pass through the focus. Conversely, if a light source is placed at the focus and the parabola is reflective, then all light rays reflected off the parabola are parallel to the axis of symmetry.
* Applications
    · The motion of the center of mass of an object under a uniform gravitational field with no air resistance is a parabola
    · If an asteroid passes by a planet at exactly the planet's escape velocity, then the asteroid's path is a parabola
    · Flashlights, telescopes, satellite dishes, etc. all use the reflective property of parabolas
    · The shape of the main cables on a suspension bridge lie between a parabola and a catenary (a catenary is a hanging cable supported at its ends)
    · Parabolas describe the surface of a liquid in a container that is rotated about a central axis

– Ellipse
    * An *ellipse* is the set of points whose sum of the distances from two distinct fixed points (called foci, plural of focus) is constant
    * The graph of an ellipse is an oval-shaped closed curve
    * Ellipses are *not* functions because they fail the vertical line test
    * Terminology
        · The line through the foci intersects the ellipse at its two *vertices* (plural of vertex)
        · The line segment between the two vertices is known as the *major axis*
        · The *center* of an ellipse is the midpoint of the major axis
        · The line segment through the center of an ellipse that has its endpoints on the ellipse and is perpendicular to the major axis is known as the *minor axis*

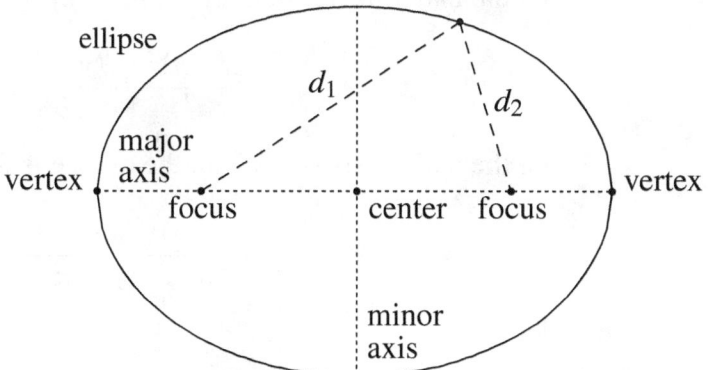

    * Parameters
        · The length of the major axis is $2a$

- The length of the minor axis is $2b$
- The distance from the center to a focus is c

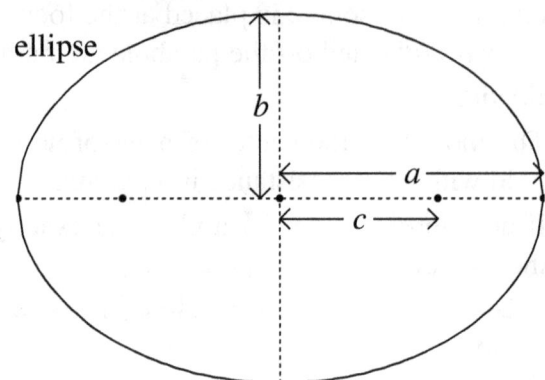

* Parameter relationships (all apply to Case I and Case II described below)
  - $a > b$
  - $c^2 = a^2 - b^2$
  - An ellipse approaches a circle as a approaches b; in the graph, the foci are approaching the center as a approaches b
* Case I: Ellipse with a center at the origin
  - When the major axis is horizontal, the ellipse is described by
  $$\frac{x^2}{a^2} + \frac{y^2}{b^2} = 1$$
  - When the major axis is vertical, the ellipse is described by
  $$\frac{x^2}{b^2} + \frac{y^2}{a^2} = 1$$
* Case II: Ellipse with a center at (h, k)
  - When the major axis is horizontal, the ellipse is described by
  $$\frac{(x-h)^2}{a^2} + \frac{(y-k)^2}{b^2} = 1$$
  - When the major axis is vertical, the ellipse is described by
  $$\frac{(x-h)^2}{b^2} + \frac{(y-k)^2}{a^2} = 1$$
* Eccentricity
  - The *eccentricity* of an ellipse is a measure of the elongation of an ellipse
  - The eccentricity of an ellipse is the ratio of the distance between the foci to the length of the major axis, that is, $2c/2a = c/a$

Chapter 4. Analytic Geometry

- The eccentricity satisfies $0 < c/a < 1$
- The graph of the ellipse tends to a circle as the eccentricity tends to 0
- The graph of the ellipse tends to a line segment (the major axis) as the eccentricity tends to 1

* Applications
  - Orbits of planets
  - Arches used in architecture
  - Pendulum motion
  - Elliptical gears
  - Modeling random variables in probability using the bivariate normal distribution
  - Light waves, sound waves, and water waves that emanate from one focus of an ellipse reflect off of the ellipse and converge to the other focus simultaneously

– Hyperbola
  * A *hyperbola* is the set of points whose difference of the distances from two distinct points (called foci) is constant
  * Terminology
    - The graph of a hyperbola consists of two disconnected *branches*
    - The line through the foci intersects the hyperbola at its *vertices*
    - The line segment connecting the vertices is called the *transverse axis*
    - The midpoint of the transverse axis is the *center* of the hyperbola

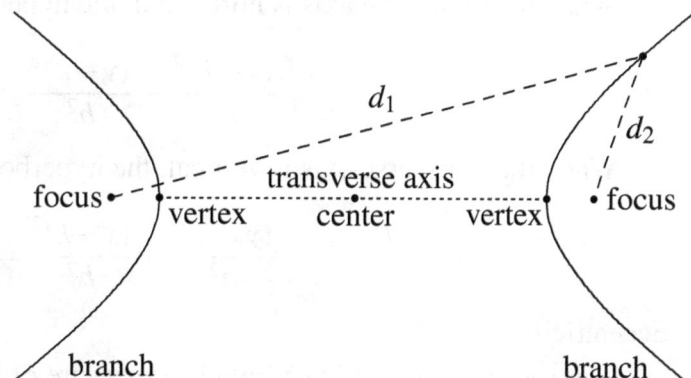

  * Hyperbolas are *not* functions because they fail the vertical line test
  * Parameters
    - The length of the transverse axis is $2a$, that is, the distance from the center to a vertex is a
    - The distance from the center to a focus is c
  * Parameter relationships (all apply to Case I and Case II described below)
    - $c^2 = a^2 + b^2$

- An $a \times b$ *reference box* is used to generate two *slant asymptotes* that intersect at the center of the hyperbola

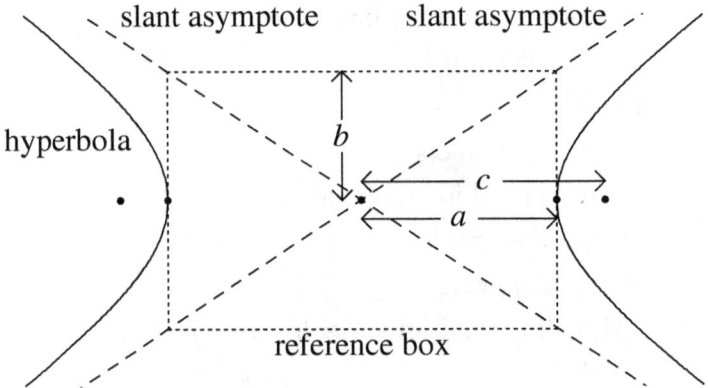

* Case I: Hyperbola with a center at the origin
  - When the transverse axis is horizontal, the hyperbola is described by
  $$\frac{x^2}{a^2} - \frac{y^2}{b^2} = 1$$
  - When the transverse axis is vertical, the hyperbola is described by
  $$\frac{y^2}{a^2} - \frac{x^2}{b^2} = 1$$

* Case II: Hyperbola with a center at (h, k)
  - When the transverse axis is horizontal, the hyperbola is described by
  $$\frac{(x-h)^2}{a^2} - \frac{(y-k)^2}{b^2} = 1$$
  - When the transverse axis is vertical, the hyperbola is described by
  $$\frac{(y-k)^2}{a^2} - \frac{(x-h)^2}{b^2} = 1$$

* Eccentricity
  - The *eccentricity* of a hyperbola is a measure of the openness (or narrowness) of the branches
  - The eccentricity of a hyperbola has the same definition as in the ellipse: c/a
  - The eccentricity satisfies $c/a > 1$
  - The branches become very narrow as the eccentricity tends to 1
  - The branches become very wide as the eccentricity tends to ∞

* Applications
  - Hyperbolic orbits

- Hyperbolic reflection: a light ray from one focus of a hyperbola reflects off the nearest branch in such a way that the reflected path lies along the line through the other focus and the point of reflection
- Sound location: if a single sound is recorded at two different times at two different locations, then the two points are the foci and the sound occurred on one branch of the hyperbola

- Direct and indirect variation

  - The statement "y is directly proportional to x" means

  $$y = kx$$

  for some nonzero constant k

  - The statement "y is inversely proportional to x" means

  $$y = k \cdot \frac{1}{x}$$

  for some nonzero constant k

  - These two senses of variation can be combined. Newton's law of universal gravitation, for example, states that the gravitational attraction F between two objects of mass m_1 and m_2 is directly proportional to the product of their masses and inversely proportional to the square of the distance r between the objects means that

  $$F = k \cdot \frac{m_1 m_2}{r^2}$$

- Parametric equations

  - Parametric equations are used to described the position of a point (x, y) in the rectangular coordinate system as a function of a third variable. The third variable is oftentimes denoted by t, for time. The parametric equations

  $$x = f(t)$$

  and

  $$y = g(t)$$

  where f and g are functions of t, describe the position of a point at time t.

  - The collection of these points in the rectangular coordinate system is known as a *plane curve*

  - The plane curve associated with parametric equations is plotted by obtaining several (x, y) pairs for various values of t, then drawing a smooth curve through the points. The plane curve has an *orientation* or direction that it follows as t increases. The orientation is often depicted with arrows along the plane curve.

- To convert from parametric equations to an equation involving just x and y, solve one of the equations for t, then substitute this into the other equation, thus eliminating t
- When trigonometric functions appear in parametric equations, an angle θ is oftentimes used instead of t
- Parametric equations are useful for describing the motion of an object over time, that is

$$(x, y) = \big(f(t), g(t)\big)$$

Two common applications are the motion of a projectile over time and the position of an object in orbit over time.

Chapter 4. Analytic Geometry

Exercises

4.1 Find $f(5)$ for the function $f(x) = 2x - 7$.

4.2 Find $f(g(2))$ for the functions $f(x) = x^2$ and $g(x) = 3x$.

4.3 Find the *zeros* of the function
$$f(x) = x^2 - 2x - 3.$$

4.4 Find the *domain* of the function
$$y = f(x) = \frac{x^2 - 1}{x^2 + 1}.$$

4.5 Find the slope of the hypotenuse of the triangle with vertices $(1, 1)$, $(4, -3)$, and $(4, 1)$.

4.6 Find the equation of the line that passes through the point $(1, 2)$ that is parallel to the line $2x - y = 7$.

4.1 Find $f(5)$ for the function $f(x) = 2x - 7$.

The function $f(x)$ evaluated at $x = 5$ is
$$f(5) = 2 \cdot 5 - 7 = 10 - 7 = 3.$$

4.2 Find $f(g(2))$ for the functions $f(x) = x^2$ and $g(x) = 3x$.

Beginning the evaluation from the inside, the composite function $f(g(x))$ evaluated at $x = 2$ is
$$f(g(2)) = f(3 \cdot 2) = f(6) = 36.$$

4.3 Find the *zeros* of the function
$$f(x) = x^2 - 2x - 3.$$

Since the quadratic function can be factored as
$$x^2 - 2x - 3 = (x+1)(x-3)$$

the zeros are $x = -1$ and $x = 3$.

4.4 Find the *domain* of the function
$$y = f(x) = \frac{x^2 - 1}{x^2 + 1}.$$

There are no restrictions on the values of x, so the domain of $f(x)$ is $\{x \mid x \in \mathbb{R}\}$.

4.5 Find the slope of the hypotenuse of the triangle with vertices $(1, 1)$, $(4, -3)$, and $(4, 1)$.

The lengths of the three sides of this triangle are 3, 4, and 5, so the hypotenuse is the longest side, which connects $(1, 1)$ and $(4, -3)$. The slope of the line connecting these two points is
$$m = \frac{1 - (-3)}{1 - 4} = -\frac{4}{3}.$$

4.6 Find the equation of the line that passes through the point $(1, 2)$ that is parallel to the line $2x - y = 7$.

Since the lines are parallel, both have slope 2. Using the point-slope formula for a line,
$$y - 2 = 2(x - 1)$$

or
$$y = 2x.$$

Chapter 4. Analytic Geometry

4.7 Which of the functions given below have an inverse function?

(a) $f(x) = \lfloor x \rfloor$ for $x \in \mathbb{R}$.

(b) $f(x) = \sqrt{9-x^2}$ for $-3 \leq x \leq 3$.

(c) $f(x) = 17$ for $x \in \mathbb{R}$.

(d) $f(x) = x$ for $x \in \mathbb{R}$.

4.8 The point $(2, -5)$ lies on the graph of the *even* function $y = f(x)$. Name a second point that lies on the graph of $y = f(x)$.

4.9 Find the inverse of the function $f(x) = 2x - 3$.

4.10 The length of each side of a cube is x. Write a function $S(x)$ which gives the surface area of the cube as a function of x.

4.7 Which of the functions given below have an inverse function?

  (a) $f(x) = \lfloor x \rfloor$ for $x \in \mathbb{R}$.
  (b) $f(x) = \sqrt{9-x^2}$ for $-3 \leq x \leq 3$.
  (c) $f(x) = 17$ for $x \in \mathbb{R}$.
  (d) $f(x) = x$ for $x \in \mathbb{R}$.

  (a) $f(x) = \lfloor x \rfloor$ for $x \in \mathbb{R}$ is a step function that does not have an inverse because it does not pass the horizontal line test.

  (b) $f(x) = \sqrt{9-x^2}$ for $-3 \leq x \leq 3$ is a semicircle that does not have an inverse because it does not pass the horizontal line test.

  (c) $f(x) = 17$ for $x \in \mathbb{R}$ is a constant function that does not have an inverse because it does not pass the horizontal line test.

  (d) $f(x) = x$ for $x \in \mathbb{R}$ has an inverse because it passes the horizontal line test.

4.8 The point $(2, -5)$ lies on the graph of the *even* function $y = f(x)$. Name a second point that lies on the graph of $y = f(x)$.

The point $(-2, -5)$ also lies on the graph because $f(-x) = f(x)$ for an even function.

4.9 Find the inverse of the function $f(x) = 2x - 3$.

Swapping x and y in the equation
$$y = 2x - 3$$
gives
$$x = 2y - 3.$$
Adding three to each side and dividing by two gives
$$y = \frac{x+3}{2}.$$
So the inverse function is
$$f^{-1}(x) = \frac{x+3}{2}.$$

4.10 The length of each side of a cube is x. Write a function $S(x)$ which gives the surface area of the cube as a function of x.

Since the cube has six sides, and each side has area x^2, the surface area as a function of x is
$$S(x) = 6x^2.$$

Chapter 4. Analytic Geometry

4.11 Which statement best describes the zeros of the cubic (third degree) polynomial function shown in the graph below?

(a) There are two real zeros and one complex zero.

(b) There are three complex zeros.

(c) There are two distinct real zeros (one with multiplicity one and the other with multiplicity two).

(d) There is one real zero (multiplicity one) and one complex zero (multiplicity two).

(e) There are no real zeros.

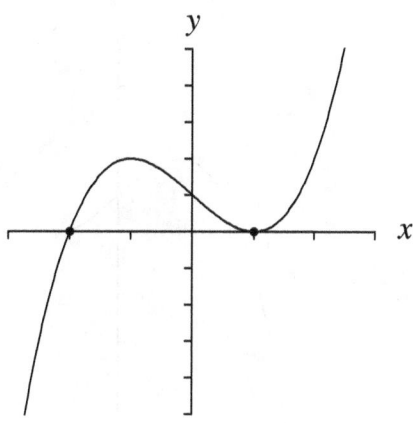

4.12 Consider the piecewise function

$$g(x) = \begin{cases} x^3 - x + 7 & x \leq 0 \\ x^2 - 3x + 36 & x > 0. \end{cases}$$

Find $g(-1)$.

4.11 Which statement best describes the zeros of the cubic (third degree) polynomial function shown in the graph below?

(a) There are two real zeros and one complex zero.

(b) There are three complex zeros.

(c) There are two distinct real zeros (one with multiplicity one and the other with multiplicity two).

(d) There is one real zero (multiplicity one) and one complex zero (multiplicity two).

(e) There are no real zeros.

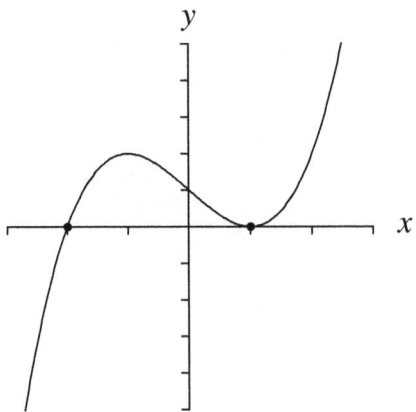

The cubic polynomial has three zeros in the complex plane. The graph indicates that the polynomial has real zeros at $x = -2$ and $x = 1$. Since the graph crosses the x-axis at $x = -2$, this root has multiplicity one. Since the graph touches, but does not cross, the x-axis at $x = 1$, this root has multiplicity two. Thus the polynomial has two distinct real zeros, one with multiplicity one and a second with multiplicity two. Response (c) best characterizes the zeros.

4.12 Consider the piecewise function

$$g(x) = \begin{cases} x^3 - x + 7 & x \leq 0 \\ x^2 - 3x + 36 & x > 0. \end{cases}$$

Find $g(-1)$.

Using the $x \leq 0$ portion of the piecewise function,

$$g(-1) = (-1)^3 - (-1) + 7 = -1 + 1 + 7 = 7.$$

Chapter 4. Analytic Geometry

4.13 Let $f(x) = \sqrt{x}$ and $g(x) = 2x + c$, for some constant c. Find the value of c so that the graph of $y = f(g(x))$ passes through the point $(4, 3)$.

4.14 A 12-inch wire is cut at a position x measured in inches from the left end of the wire, where $0 < x < 12$. The left-hand piece of the wire is bent into a circle and the right-hand piece is bent into a square. Write the sum of the areas of the circle and the square as a function of x and determine the value of x that minimizes the sum of the areas.

4.15 Consider the functions
$$f(x) = x^3 + 3x^2 - x + 7$$
and
$$g(x) = x^3 + 9x^2 + 23x + 25.$$
Which one of the following eight formulas describes the relationship between $f(x)$ and $g(x)$?

$$g(x) = 2f(x) \qquad g(x) = f(x)/2 \qquad g(x) = f(2x) \qquad g(x) = f(x/2)$$
$$g(x) = f(x) + 2 \qquad g(x) = f(x) - 2 \qquad g(x) = f(x+2) \qquad g(x) = f(x-2).$$

4.13 Let $f(x) = \sqrt{x}$ and $g(x) = 2x + c$, for some constant c. Find the value of c so that the graph of $y = f(g(x))$ passes through the point $(4, 3)$.

Since $y = f(g(x)) = f(2x + c) = \sqrt{2x + c}$ and the point $(4, 3)$ is on the graph, the constant c must satisfy
$$3 = \sqrt{2 \cdot 4 + c}.$$
Solving this equation for c yields $c = 1$.

4.14 A 12-inch wire is cut at a position x measured in inches from the left end of the wire, where $0 < x < 12$. The left-hand piece of the wire is bent into a circle and the right-hand piece is bent into a square. Write the sum of the areas of the circle and the square as a function of x and determine the value of x that minimizes the sum of the areas.

The length of the left-hand piece of wire is x; the length of the right-hand piece of wire is $12 - x$. Let r be the radius of the circle and s be the length of a side of the square. The circumference of the circle is $2\pi r = x$, so the radius of the circle is $r = x/(2\pi)$. The perimeter of the square is $4s = 12 - x$, so the length of a side of the square is $s = (12 - x)/4$. Using the formulas for the area of a circle and the area of a square, the sum of the two areas as a function of x is

$$a(x) = \pi \left(\frac{x}{2\pi}\right)^2 + \left(\frac{12-x}{4}\right)^2 = \frac{x^2}{4\pi} + \frac{144 - 24x + x^2}{16} = \left(\frac{1}{16} + \frac{1}{4\pi}\right)x^2 - \frac{3}{2}x + 9.$$

Since the leading coefficient of this quadratic function is positive, its graph is a concave-up parabola with a vertex at the x-value

$$x = -\frac{b}{2a} = \frac{\frac{3}{2}}{\frac{1}{8} + \frac{1}{2\pi}} = \frac{12\pi}{\pi + 4} \cong 5.2788,$$

which is the value of x that minimizes the sum of the areas of the circle and the square.

4.15 Consider the functions
$$f(x) = x^3 + 3x^2 - x + 7$$
and
$$g(x) = x^3 + 9x^2 + 23x + 25.$$
Which one of the following eight formulas describes the relationship between $f(x)$ and $g(x)$?

$g(x) = 2f(x)$ $\quad$ $g(x) = f(x)/2$ $\quad$ $g(x) = f(2x)$ $\quad$ $g(x) = f(x/2)$

$g(x) = f(x) + 2$ $\quad$ $g(x) = f(x) - 2$ $\quad$ $g(x) = f(x + 2)$ $\quad$ $g(x) = f(x - 2)$.

Because the leading coefficients of $f(x)$ and $g(x)$ are the same, the first four relationships can be eliminated. Because $f(x)$ and $g(x)$ do not differ only by 2 in the constant term, the next two relationships can be eliminated. It must be one of the last two, which correspond to a horizontal shift in the graph. Using a test point of $x = 0$, $g(0) = f(2) = 25$, but $g(0) \neq f(-2)$, so the correct relationship is $g(x) = f(x + 2)$.

Chapter 4. Analytic Geometry

4.16 If $(x+2)$ and $(x-5)$ are factors of the function

$$f(x) = x^4 - 5x^3 - 7x^2 + 29x + 30,$$

what are the other two factors?

4.17 List the quadrant(s) in which the graph of the following equations are located.

(a) $(x+4)^2 + (y+10)^2 = 25$

(b) $|x| + |y| = 17$

(c) $xy = -17$

4.18 Consider the following five functions defined on the domain $\{x \mid x \in \mathbb{R}, 0 < x < 1\}$.

(a) $f(x) = x^5$

(b) $f(x) = |x|$

(c) $f(x) = x^2$

(d) $f(x) = 1/x$

(e) $f(x) = \sqrt{x}$

Which function does *not* have range $\{y \mid y \in \mathbb{R}, 0 < y < 1\}$?

4.16 If $(x+2)$ and $(x-5)$ are factors of the function

$$f(x) = x^4 - 5x^3 - 7x^2 + 29x + 30,$$

what are the other two factors?

Using synthetic division with -2, then with 5, gives the quadratic equation $x^2 - 2x - 3$, so

$$x^4 - 5x^3 - 7x^2 + 29x + 30 = (x+2)(x-5)\left(x^2 - 2x - 3\right).$$

Factoring the quadratic equation gives

$$x^4 - 5x^3 - 7x^2 + 29x + 30 = (x+2)(x-5)\left(x^2 - 2x - 3\right) = (x+2)(x-5)(x+1)(x-3),$$

so the other two factors are $(x+1)$ and $(x-3)$.

4.17 List the quadrant(s) in which the graph of the following equations are located.

(a) $(x+4)^2 + (y+10)^2 = 25$

(b) $|x| + |y| = 17$

(c) $xy = -17$

(a) The graph of the equation $(x+4)^2 + (y+10)^2 = 25$ is a circle of radius 5 centered at $(-4, -10)$. A sketch of this circle reveals that the circle is in the third and fourth quadrants.

(b) The graph of the equation $|x| + |y| = 17$ is a diamond-shaped figure (that is, a square rotated $45°$) centered at the origin. It is located in all four quadrants.

(c) The graph of the function $y = -17/x$, for all $x \neq 0$, always has opposite signs for x and y, so its graph is in the second and fourth quadrants.

4.18 Consider the following five functions defined on the domain $\{x \,|\, x \in \mathbb{R}, 0 < x < 1\}$.

(a) $f(x) = x^5$

(b) $f(x) = |x|$

(c) $f(x) = x^2$

(d) $f(x) = 1/x$

(e) $f(x) = \sqrt{x}$

Which function does *not* have range $\{y \,|\, y \in \mathbb{R}, 0 < y < 1\}$?

The function $f(x) = 1/x$ has range $\{y \,|\, y \in \mathbb{R}, 1 < y < \infty\}$, so the correct response is (d).

Chapter 4. Analytic Geometry

4.19 Write the equations of all of the asymptotes associated with the rational function

$$y = \frac{4x^2 - 7}{x(x-3)}.$$

4.20 Write the equation of a cubic polynomial function in standard form that has the following zeros:

$$-2, 0, 3.$$

4.21 The dollar amount A after t years of continuous compounding interest on an initial principal amount P at an annual interest rate r is

$$A = Pe^{rt}.$$

Give a formula for the amount of time it takes to double an initial principal amount P as a function of the annual interest rate r.

4.19 Write the equations of all of the asymptotes associated with the rational function
$$y = \frac{4x^2 - 7}{x(x-3)}.$$

The rational function has no factors common to both the numerator and denominator. The zeros of the polynomial in the denominator correspond to vertical asymptotes, so there will be vertical asymptotes at $x = 0$ and $x = 3$. Since the degrees of the polynomials in the numerator and denominator match, there will be a horizontal asymptote at the ratio of the leading coefficients, which is at $y = 4$.

4.20 Write the equation of a cubic polynomial function in standard form that has the following zeros:
$$-2, 0, 3.$$

One such polynomial function is
$$\begin{aligned} f(x) &= (x+2)(x-0)(x-3) \\ &= x(x^2 - x - 6) \\ &= x^3 - x^2 - 6x. \end{aligned}$$

The polynomial function $g(x) = 17(x^3 - x^2 - 6x)$ also has zeros $-2, 0, 3$.

4.21 The dollar amount A after t years of continuous compounding interest on an initial principal amount P at an annual interest rate r is
$$A = Pe^{rt}.$$

Give a formula for the amount of time it takes to double an initial principal amount P as a function of the annual interest rate r.

The equation
$$2P = Pe^{rt}$$
needs to be solved for t in order to find the doubling time. Dividing the equation by P gives
$$2 = e^{rt}.$$

Taking the natural logarithm of both sides of this equation gives
$$\ln 2 = \ln e^{rt}$$

or
$$\ln 2 = rt.$$

Dividing both sides of this equation by r gives the doubling time as
$$t = \frac{\ln 2}{r}.$$

Chapter 4. Analytic Geometry

4.22 Find the equations of the asymptotes of the hyperbola associated with the equation

$$x^2 - 4y^2 = 36.$$

4.23 The graph of the function $y = f(x)$ is shown below.

(a) Write the function $y = f(x)$ using the absolute value function.

(b) Find $f(f(1))$.

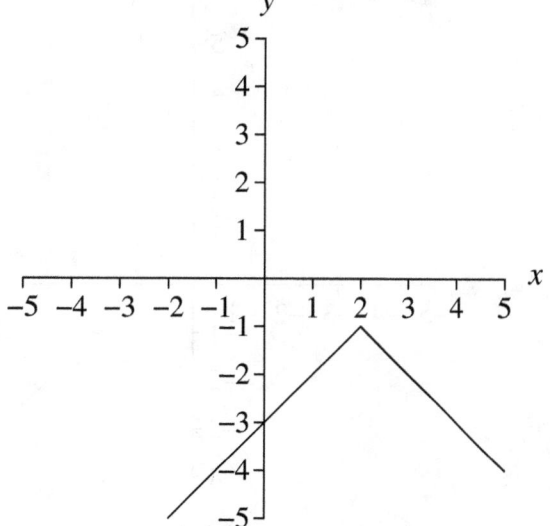

4.22 Find the equations of the asymptotes of the hyperbola associated with the equation
$$x^2 - 4y^2 = 36.$$

Dividing both sides of the equation by 36 places the equation in standard form:
$$\frac{x^2}{36} - \frac{y^2}{9} = 1.$$

This equation corresponds to a hyperbola that is centered at the origin with a horizontal transverse axis and $a = 6$ and $b = 3$. The slant asymptotes run through the opposite corners of the reference box, which is $2a = 12$ units wide and $2b = 6$ units high. Using the "rise over run" definition of the slope, the vertical asymptotes are
$$y = \pm \frac{1}{2}x.$$

4.23 The graph of the function $y = f(x)$ is shown below.

(a) Write the function $y = f(x)$ using the absolute value function.
(b) Find $f(f(1))$.

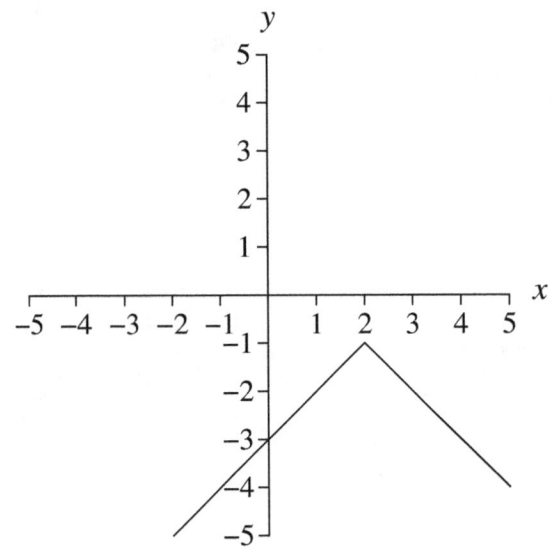

(a) Beginning with the parent function $y = |x|$, the following transformations result in an equation for the function.
- a horizontal shift two units to the right: $y = |x - 2|$
- a reflection across the x axis: $y = -|x - 2|$
- a vertical shift down one unit: $y = -|x - 2| - 1$

(b) The value of $f(f(1))$ is
$$f(f(1)) = f(-2) = -5.$$

4.24 Find the *x*-intercept(s) and *y*-intercept(s) of the graph of the equation
$$\frac{(x-3)^2}{9} + \frac{(y-2)^2}{16} = 1.$$
Express your solutions as ordered pairs.

4.24 Find the x-intercept(s) and y-intercept(s) of the graph of the equation

$$\frac{(x-3)^2}{9} + \frac{(y-2)^2}{16} = 1.$$

Express your solutions as ordered pairs.

The equation corresponds to an ellipse centered at $(3, 2)$ with a vertical major axis of length $2a = 8$ and a minor axis of length $2b = 6$. The x-intercepts can be found by setting y to 0 and solving for x, which gives

$$\frac{(x-3)^2}{9} + \frac{(0-2)^2}{16} = 1$$

or

$$\frac{(x-3)^2}{9} = \frac{3}{4}$$

or

$$(x-3)^2 = \frac{27}{4}$$

or

$$x - 3 = \sqrt{\frac{27}{4}} \quad \text{or} \quad x - 3 = -\sqrt{\frac{27}{4}}$$

or

$$x = 3 + \frac{\sqrt{27}}{2} \quad \text{or} \quad x = 3 - \frac{\sqrt{27}}{2}.$$

So the x-intercepts are the ordered pairs

$$\left(3 - \frac{\sqrt{27}}{2}, 0\right) \quad \text{and} \quad \left(3 + \frac{\sqrt{27}}{2}, 0\right).$$

To find the y-intercepts, set x to 0 and solve for y, which gives

$$\frac{(0-3)^2}{9} + \frac{(y-2)^2}{16} = 1$$

or

$$\frac{(y-2)^2}{16} = 0$$

or

$$(y-2)^2 = 0$$

or

$$y - 2 = 0$$

or

$$y = 2.$$

So there is a single y-intercept at $(0, 2)$.

Chapter 4. Analytic Geometry

4.25 Find the *range* of the function
$$y = f(x) = \frac{x^2 - 1}{x^2 + 1}.$$

4.26 Find the function of the form
$$y = f(x) = ax^2 + bx + c$$
that passes through the points $(1, 4)$, $(-1, -2)$, and $(0, 0)$.

4.25 Find the *range* of the function
$$y = f(x) = \frac{x^2-1}{x^2+1}.$$

The rational function can be written as
$$y = f(x) = \frac{x^2-1}{x^2+1} = \frac{x^2+1-2}{x^2+1} = 1 - \frac{2}{x^2+1},$$

which is a continuous even function that is increasing on $[0, \infty)$. Since $f(0) = -1$ and $f(x)$ approaches 1 as $x \to \infty$, the range of $f(x)$ is the interval $[-1, 1)$, which is closed on the left and open on the right. Plotting points to get the graph of $y = f(x)$ would also lead to this conclusion.

4.26 Find the function of the form
$$y = f(x) = ax^2 + bx + c$$

that passes through the points $(1, 4)$, $(-1, -2)$, and $(0, 0)$.

Plugging the three points into the quadratic equation gives the following set of three equations with three unknowns:
$$a + b + c = 4$$
$$a - b + c = -2$$
$$c = 0.$$

Plugging $c = 0$ into the first two equations reduces the set of equations to two equations and two unknowns:
$$a + b = 4$$
$$a - b = -2.$$

Solving the second equation for a gives $a = b - 2$. Substituting this value of a into the first equation gives
$$(b - 2) + b = 4$$
or
$$2b - 2 = 4.$$

Adding 2 to both sides of this equation, then dividing both sides of the resulting equation by 2 gives
$$b = 3.$$

Since $a = b - 2$, the value of a must be $a = 1$. So finally the quadratic function
$$y = f(x) = x^2 + 3x$$

passes through the points $(1, 4)$, $(-1, -2)$, and $(0, 0)$.

Chapter 4. Analytic Geometry

4.27 Find the equation of all points (x, y) such that the distance between (x, y) and $(2, 3)$ equals the distance between (x, y) and $(6, -3)$.

4.28 A farmer has 100 feet of fencing. He would like to use the fencing to enclose three sides of a rectangular-shaped pig pen to be built against a wall of the barn; that is, the barn is being used as one side of the enclosure. Find the length of the side of the pen parallel to the barn x that gives the largest area of the pen.

4.27 Find the equation of all points (x, y) such that the distance between (x, y) and $(2, 3)$ equals the distance between (x, y) and $(6, -3)$.

Using the distance formula and equating the two distances referred to in the problem:
$$\sqrt{(x-2)^2 + (y-3)^2} = \sqrt{(x-6)^2 + (y+3)^2}.$$

Squaring both sides and expanding gives
$$x^2 - 4x + 4 + y^2 - 6y + 9 = x^2 - 12x + 36 + y^2 + 6y + 9$$

which simplifies to
$$8x - 12y = 32$$

or
$$2x - 3y = 8.$$

Rearranging terms so as to put this in the slope–intercept form for a line:
$$y = \frac{2}{3}x - \frac{8}{3}.$$

4.28 A farmer has 100 feet of fencing. He would like to use the fencing to enclose three sides of a rectangular-shaped pig pen to be built against a wall of the barn; that is, the barn is being used as one side of the enclosure. Find the length of the side of the pen parallel to the barn x that gives the largest area of the pen.

Let x be the length of fencing devoted to the side parallel to the barn. This leaves $100 - x$ feet of fencing for the other two sides. Since the pig pen is rectangular shaped, this means each of the two sides perpendicular to the barn has length $(100 - x)/2$. So the area of the pig pen as a function of x is just the product of the lengths of the two sides:
$$A(x) = x\left(\frac{100-x}{2}\right)$$

Simplifying and completing the square,
$$\begin{aligned} A(x) &= x\left(\frac{100-x}{2}\right) \\ &= \frac{1}{2}(100x - x^2) \\ &= -\frac{1}{2}(x^2 - 100x) \\ &= -\frac{1}{2}(x^2 - 100x + 2500) + 1250 \\ &= -\frac{1}{2}(x - 50)^2 + 1250. \end{aligned}$$

The graph of this function is a concave-down parabola with vertex $(50, 1250)$, so the optimal value of x is $x = 50$ which will give the pigs 1250 square feet of grazing space.

4.29 The graph of the piecewise linear function $y = f(x)$, which is defined on the domain $\{x \mid x \in \mathbb{R},\ 1 \leq x < 4\}$, is shown below.

(a) Find $f(2)$.

(b) Find $f(f(3))$.

(c) Find the range of the function $f(x)$.

(d) Write $f(x)$ using the notation for a piecewise function.

(e) Is $f(x)$ a one-to-one function?

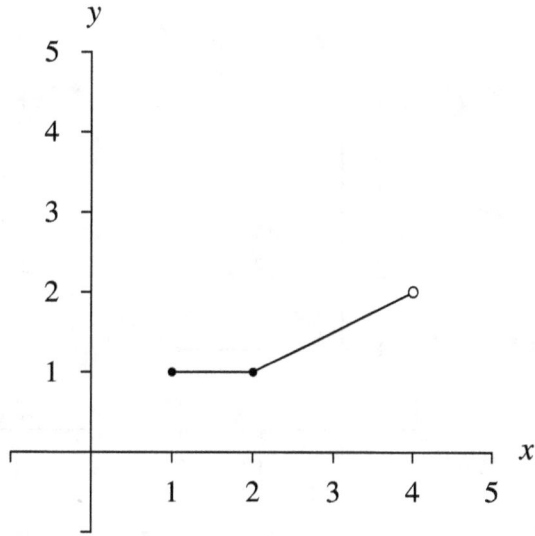

4.29 The graph of the piecewise linear function $y = f(x)$, which is defined on the domain $\{x \,|\, x \in \mathbb{R},\, 1 \leq x < 4\}$, is shown below.

(a) Find $f(2)$.

(b) Find $f(f(3))$.

(c) Find the range of the function $f(x)$.

(d) Write $f(x)$ using the notation for a piecewise function.

(e) Is $f(x)$ a one-to-one function?

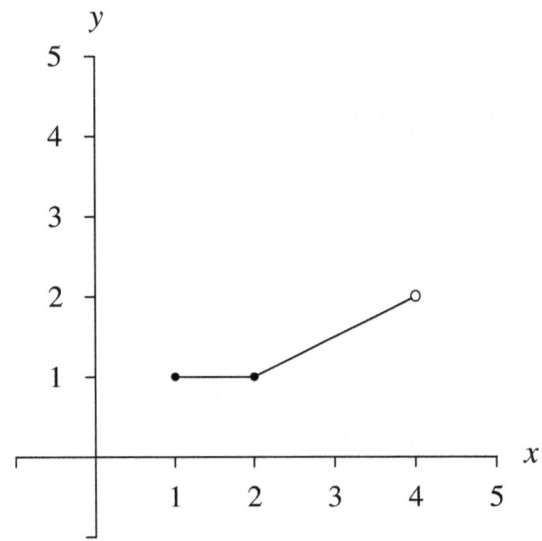

(a) The height of $y = f(x)$ at $x = 2$ is 1, so $f(2) = 1$.

(b) The height of the function at $x = 3$ is 1.5. Evaluating $f(x)$ at $x = 1.5$ gives 1, so $f(f(3)) = f(1.5) = 1$.

(c) The range of $y = f(x)$ is the interval $[1, 2)$. This interval can also be written as the inequality $1 \leq y < 2$ or expressed as the set
$$\{y \,|\, y \in \mathbb{R},\, 1 \leq y < 2\}.$$

(d) Using the notation for a piecewise function,
$$f(x) = \begin{cases} 1 & 1 \leq x \leq 2 \\ x/2 & 2 < x < 4 \end{cases}$$
because $f(x)$ is constant for $1 \leq x \leq 2$ and $f(x)$ is a line with slope $m = 1/2$ and y-intercept 0 for $2 < x < 4$. (The $x = 2$ value can be allocated to either piece.)

(e) The function $f(x)$ is *not* a one-to-one function because it fails the horizontal line test at $y = 1$.

4.30 The graph of the piecewise linear function $y = f(x)$ is shown on the left and the graph of the piecewise linear function $y = g(x)$ is shown on the right.

(a) Find $f(2) + g(1)$.

(b) Find $g(f(2))$.

(c) Find $f^{-1}(g(2))$.

(d) Find the domain of $f(x)/g(x)$.

(e) Using three rigid transformations, write $g(x)$ in terms of f and x.

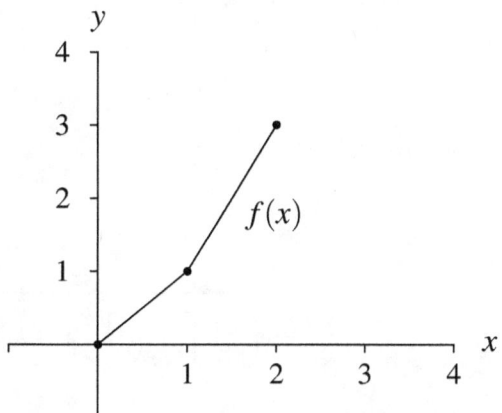
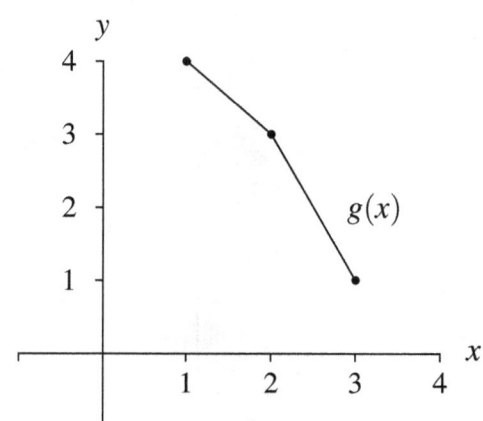

4.30 The graph of the piecewise linear function $y = f(x)$ is shown on the left and the graph of the piecewise linear function $y = g(x)$ is shown on the right.

(a) Find $f(2) + g(1)$.
(b) Find $g(f(2))$.
(c) Find $f^{-1}(g(2))$.
(d) Find the domain of $f(x)/g(x)$.
(e) Using three rigid transformations, write $g(x)$ in terms of f and x.

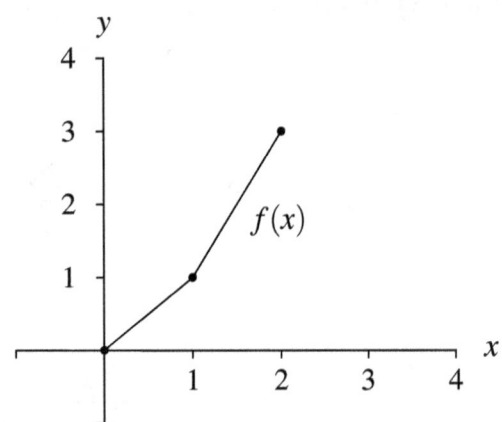
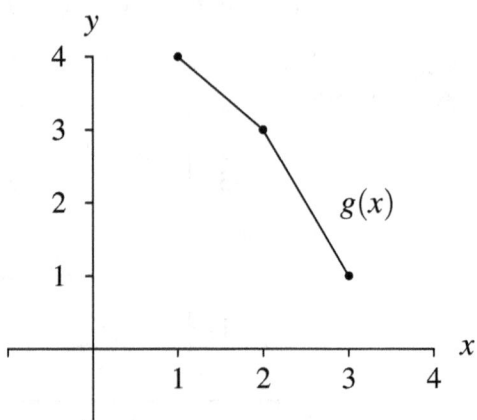

(a) The sum of $f(2)$ and $g(1)$ is
$$f(2) + g(1) = 3 + 4 = 7.$$

(b) The composition of g and f evaluated at $x = 2$ is
$$g(f(2)) = g(3) = 1.$$

(c) The composition of f^{-1} and g evaluated at $x = 2$ is
$$f^{-1}(g(2)) = f^{-1}(3) = 2.$$

(d) The domain of the function $f(x)/g(x)$ is the intersection of the domain of $f(x)$, which is $\{x \,|\, x \in \mathbb{R}, 0 \leq x \leq 2\}$, and the domain of $g(x)$, which is $\{x \,|\, x \in \mathbb{R}, 1 \leq x \leq 3\}$. The domain of $f(x)/g(x)$ is
$$\{x \,|\, x \in \mathbb{R}, 1 \leq x \leq 2\}.$$

(e) In order to write $g(x)$ in terms of f and x, use the following rigid transformations (in order):

- shift the graph of $f(x)$ one unit to the right,
- rotate (reflect) this graph about the x-axis,
- shift this graph up four units.

Thus, $g(x)$ can be written in terms of $f(x)$ and x as
$$g(x) = -f(x-1) + 4.$$

Chapter 5

Trigonometry

Trigonometry is the study of triangles, with an emphasis on the lengths of their sides and the angles between their sides. Trigonometric functions, such as the sine, cosine, and tangent, are functions of angles that can be used to model cyclic phenomena.

- An *angle* can be formed by rotating a ray about its vertex
    - The original position of the ray is known as the *initial side* of the angle
    - The final position of the ray is known as the *terminal side* of the angle
    - Positive-valued angles are measured in the counterclockwise direction
    - Negative-valued angles are measured in the clockwise direction
    - Greek letters (for example, α, β, or θ) are often used to denote the measure of an angle
    - An angle is in its *standard position* when the initial side points eastward

    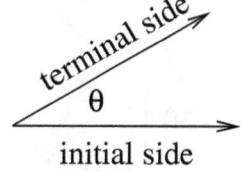

    - Angles can be measured in *degrees* and *radians*
        * Degrees are denoted by a circular superscript
        * A full rotation is 360°
        * A full rotation is 2π radians
        * To convert from degrees to radians, multiply by the unit multiplier
        $$\frac{\pi}{180°}$$
        * To convert from radians to degrees, multiply by the unit multiplier
        $$\frac{180°}{\pi}$$

- A third, less common, way to measure angles is in *grads*
- *Coterminal angles* have the same initial side and the same terminal side, for example, the angles 30°, 390°, and −330° are coterminal
- A *central angle* of a circle has its vertex at the center of a circle
- The *radian measure* of a central angle equals the associated arclength measured in radius units measured counterclockwise from the initial side of an angle, that is, the length of the associated arc divided by the length of the radius

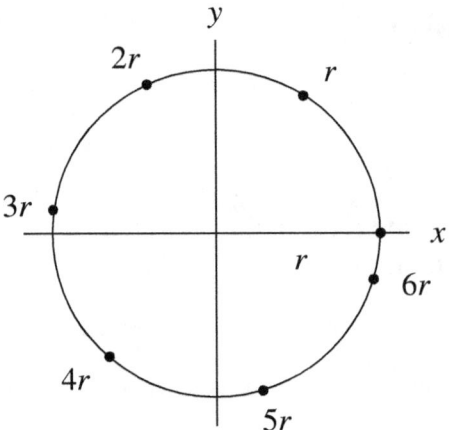

- For a 30°, 60°, 90° triangle, the lengths of the two legs are x and $\sqrt{3}x$, and the length of the hypotenuse is $2x$, where x can be any positive real number

- For a 45°, 45°, 90° triangle, the lengths of the two legs are both x and the length of the hypotenuse is $\sqrt{2}x$, where x can be any positive real number

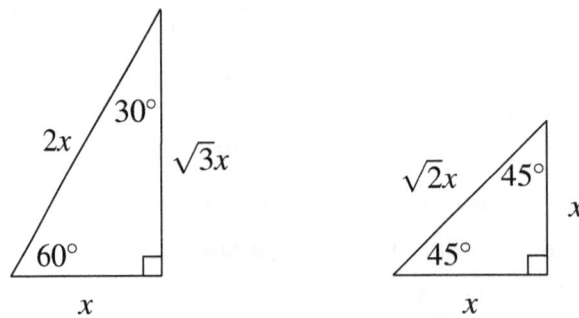

- Trigonometric functions for a right triangle with acute angle θ

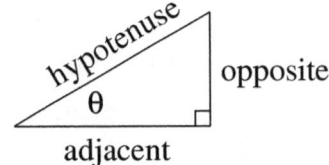

Chapter 5. Trigonometry

- The mnemonic "SOH CAH TOA" is helpful for remembering the definitions of the sin ("sine"), cos ("cosine"), and tan ("tangent") functions of an angle θ in a right triangle, which are

$$\sin\theta = \frac{\text{opposite}}{\text{hypotenuse}}$$

$$\cos\theta = \frac{\text{adjacent}}{\text{hypotenuse}}$$

$$\tan\theta = \frac{\text{opposite}}{\text{adjacent}}$$

The trigonometric functions are abbreviated by their first three letters: sin, cos, tan.

- The other three trigonometric functions (cosecant, secant, and cotangent) can be defined as reciprocals of the others:

$$\csc\theta = \frac{1}{\sin\theta} = \frac{\text{hypotenuse}}{\text{opposite}}$$

$$\sec\theta = \frac{1}{\cos\theta} = \frac{\text{hypotenuse}}{\text{adjacent}}$$

$$\cot\theta = \frac{1}{\tan\theta} = \frac{\text{adjacent}}{\text{opposite}}$$

- Cofunctions (functions that are equal for arguments that are complementary angles)
    * The sine and cosine functions are cofunctions
    * The tangent and cotangent functions are cofunctions
    * The secant and cosecant functions are cofunctions

- Winding function. Defining the six trigonometric functions in terms of triangles limits θ to $0 < \theta < 90°$. A more general approach uses the "winding function" which associates the coordinates of any point on a unit circle (a circle with radius 1 centered at the origin) on a Cartesian plane with $(\cos\theta, \sin\theta)$, where θ is the angle measured counterclockwise from the positive x-axis.

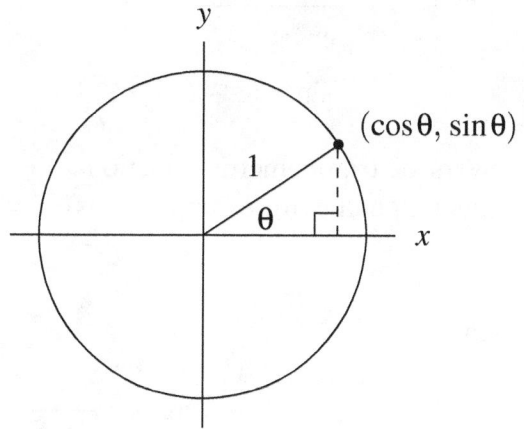

– Winding function for angles that are multiples of 30° and 45°

degrees	0°	30°	45°	60°	90°	120°	135°	150°	180°	210°	⋯
radians	0	$\frac{\pi}{6}$	$\frac{\pi}{4}$	$\frac{\pi}{3}$	$\frac{\pi}{2}$	$\frac{2\pi}{3}$	$\frac{3\pi}{4}$	$\frac{5\pi}{6}$	π	$\frac{7\pi}{6}$	⋯
$\cos\theta$	1	$\frac{\sqrt{3}}{2}$	$\frac{1}{\sqrt{2}}$	$\frac{1}{2}$	0	$-\frac{1}{2}$	$-\frac{1}{\sqrt{2}}$	$-\frac{\sqrt{3}}{2}$	-1	$-\frac{\sqrt{3}}{2}$	⋯
$\sin\theta$	0	$\frac{1}{2}$	$\frac{1}{\sqrt{2}}$	$\frac{\sqrt{3}}{2}$	1	$\frac{\sqrt{3}}{2}$	$\frac{1}{\sqrt{2}}$	$\frac{1}{2}$	0	$-\frac{1}{2}$	⋯

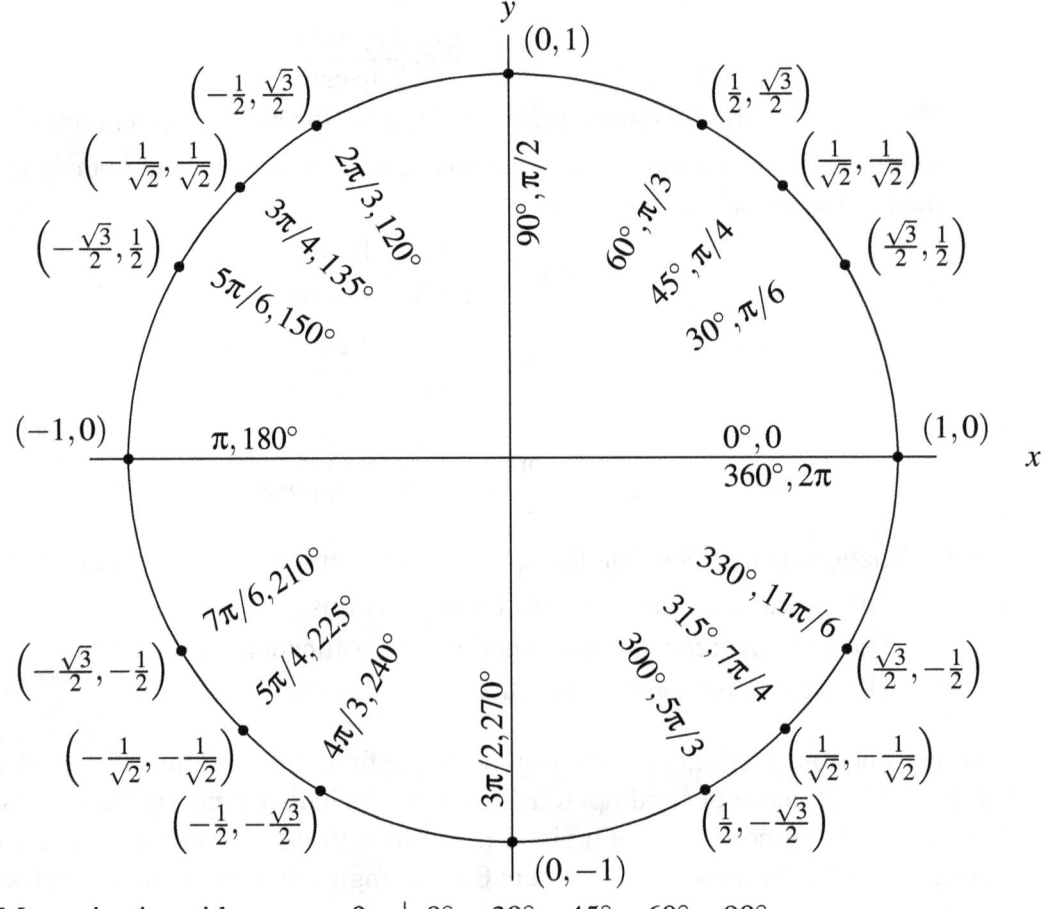

– Memorization aid:

θ	0°	30°	45°	60°	90°
$\cos\theta$	$\frac{\sqrt{4}}{2}$	$\frac{\sqrt{3}}{2}$	$\frac{\sqrt{2}}{2}$	$\frac{\sqrt{1}}{2}$	$\frac{\sqrt{0}}{2}$
$\sin\theta$	$\frac{\sqrt{0}}{2}$	$\frac{\sqrt{1}}{2}$	$\frac{\sqrt{2}}{2}$	$\frac{\sqrt{3}}{2}$	$\frac{\sqrt{4}}{2}$

- By convention, powers of trigonometric functions are written more simply by placing a superscript on the function name, for example, $(\cos\theta)^3 = \cos^3\theta$

- Trigonometric identities

    – Reciprocal identities

$$\sin\theta = \frac{1}{\csc\theta} \qquad \cos\theta = \frac{1}{\sec\theta} \qquad \tan\theta = \frac{1}{\cot\theta}$$

Chapter 5. Trigonometry

$$\csc\theta = \frac{1}{\sin\theta} \qquad \sec\theta = \frac{1}{\cos\theta} \qquad \cot\theta = \frac{1}{\tan\theta}$$

- Quotient identities

$$\tan\theta = \frac{\sin\theta}{\cos\theta} \qquad \cot\theta = \frac{\cos\theta}{\sin\theta}$$

- Pythagorean identities
  * The fundamental Pythagorean identity is

  $$\sin^2\theta + \cos^2\theta = 1$$

  * Other Pythagorean identities can be derived from the fundamental trigonometric identity; for example, dividing both sides of the fundamental trigonometric identity by $\cos^2\theta$ yields

  $$\tan^2\theta + 1 = \sec^2\theta$$

  or dividing both sides of the fundamental trigonometric identity by $\sin^2\theta$ yields

  $$1 + \cot^2\theta = \csc^2\theta$$

- Complementary angle identities (stated with radian measure)

$$\sin\left(\frac{\pi}{2} - \theta\right) = \cos\theta \qquad \cos\left(\frac{\pi}{2} - \theta\right) = \sin\theta$$

$$\tan\left(\frac{\pi}{2} - \theta\right) = \cot\theta \qquad \cot\left(\frac{\pi}{2} - \theta\right) = \tan\theta$$

$$\sec\left(\frac{\pi}{2} - \theta\right) = \csc\theta \qquad \csc\left(\frac{\pi}{2} - \theta\right) = \sec\theta$$

- Even/odd identities

$$\sin(-\theta) = -\sin\theta \qquad \cos(-\theta) = \cos\theta$$

$$\tan(-\theta) = -\tan\theta \qquad \cot(-\theta) = -\cot\theta$$

$$\sec(-\theta) = \sec\theta \qquad \csc(-\theta) = -\csc\theta$$

- Graphs of trigonometric functions (x measured in radians) have an infinite number of replications to the left and to the right. The width of each replication is known as the *period*.

function name	domain (for any integer k)	range	even/odd
$y = f(x) = \sin x$	$-\infty < x < \infty$	$-1 \leq y \leq 1$	odd
$y = f(x) = \cos x$	$-\infty < x < \infty$	$-1 \leq y \leq 1$	even
$y = f(x) = \tan x$	$-\infty < x < \infty;\ x \neq (2k+1)\pi/2$	$-\infty < y < \infty$	odd
$y = f(x) = \cot x$	$-\infty < x < \infty;\ x \neq k\pi$	$-\infty < y < \infty$	odd
$y = f(x) = \sec x$	$-\infty < x < \infty;\ x \neq (2k+1)\pi/2$	$y \leq -1$ or $y \geq 1$	even
$y = f(x) = \csc x$	$-\infty < x < \infty;\ x \neq k\pi$	$y \leq -1$ or $y \geq 1$	odd

- The sine function has period 2π

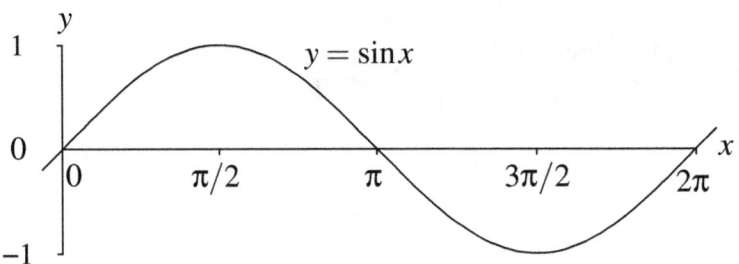

- The cosine function also has period 2π

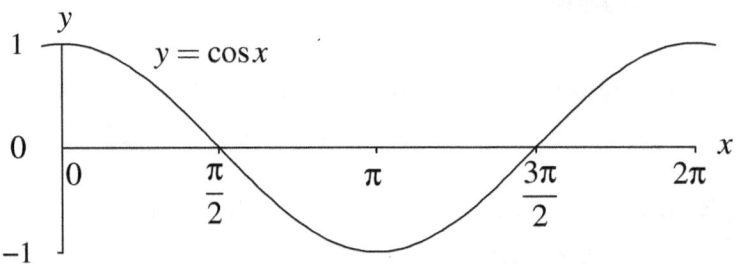

- The tangent function has period π with vertical asymptotes at $\pm\dfrac{\pi}{2}, \pm\dfrac{3\pi}{2}, \ldots$

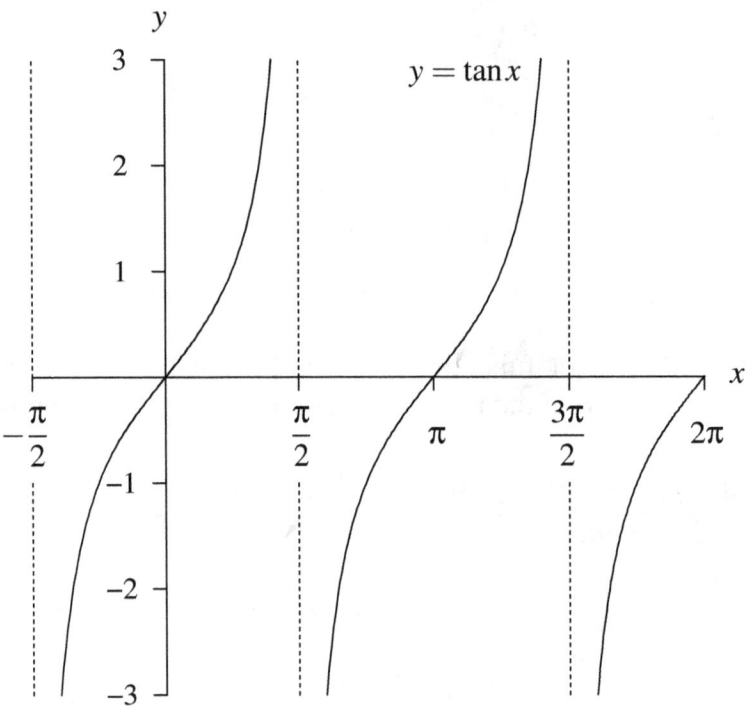

- The cotangent function has period π with vertical asymptotes at $0, \pm\pi, \pm 2\pi, \ldots$

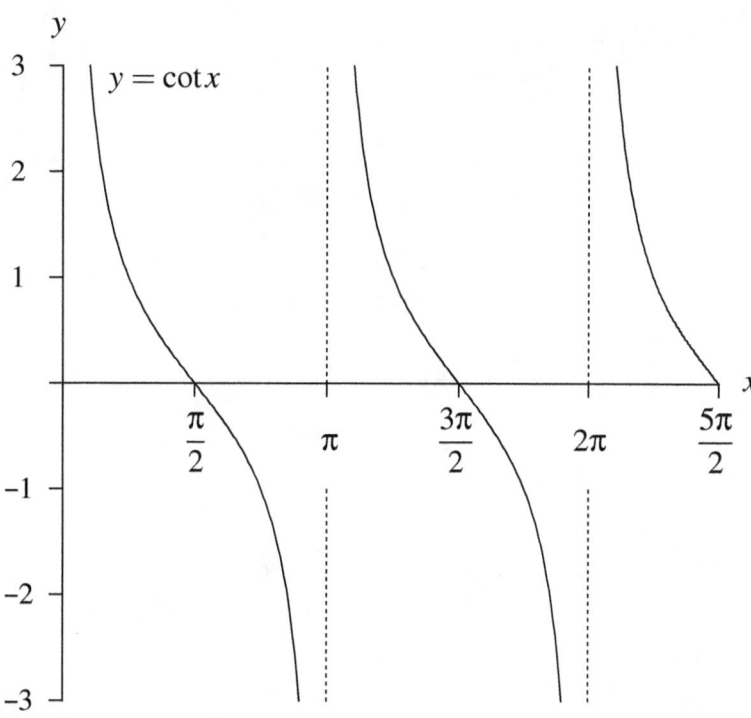

- The secant function has period 2π with vertical asymptotes at $\pm\dfrac{\pi}{2}, \pm\dfrac{3\pi}{2}, \ldots$

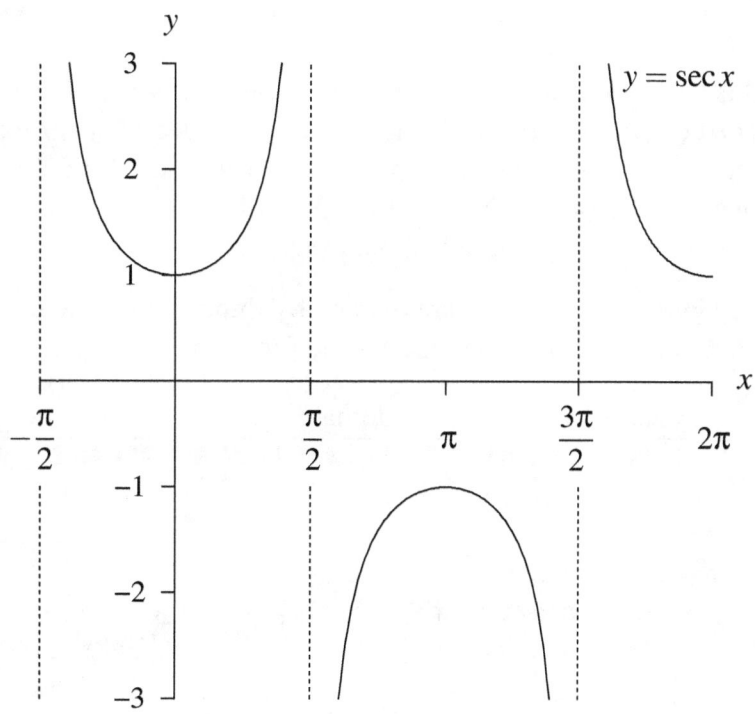

– The cosecant function has period 2π with vertical asymptotes at $0, \pm\pi, \pm 2\pi, \ldots$

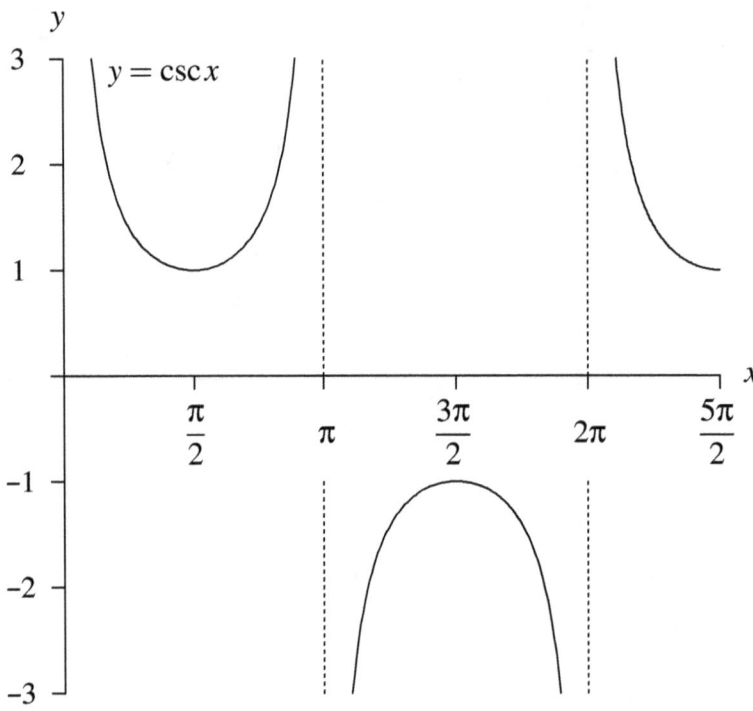

- The graphs of the trigonometric functions can be modified, for example,

$$y = a\cos(bx - c)$$

where $|a|$ is the *amplitude*, $2\pi/b$ is the *period*, and c/b is the *phase angle* or *phase shift*. The parameter b corresponds to a horizontal stretch ($0 < b < 1$) or shrink ($b > 1$); the parameter c corresponds to a horizontal shift right ($c > 0$) or left ($c < 0$); the parameter $|a|$ corresponds to a vertical stretch ($|a| > 1$) or shrink ($0 < |a| < 1$); negative values of the parameter a correspond to reflecting the graph about the x axis.

- *Inverse trigonometric functions* are defined by limiting the domains of the standard trigonometric functions so that they are one-to-one functions.

function name	domain	range
$y = f(x) = \arcsin x$	$-1 \leq x \leq 1$	$-\pi/2 \leq y \leq \pi/2$
$y = f(x) = \arccos x$	$-1 \leq x \leq 1$	$0 \leq y \leq \pi$
$y = f(x) = \arctan x$	$-\infty < x < \infty$	$-\pi/2 < y < \pi/2$
$y = f(x) = \text{arccot}\, x$	$-\infty < x < \infty$	$0 < y < \pi$
$y = f(x) = \text{arcsec}\, x$	$x \leq -1$ or $x \geq 1$	$0 \leq y < \pi/2$ or $\pi/2 < y \leq \pi$
$y = f(x) = \text{arccsc}\, x$	$x \leq -1$ or $x \geq 1$	$-\pi/2 \leq y < 0$ or $0 < y \leq \pi/2$

The function names $\arcsin x$ and $\sin^{-1} x$, for example, are used interchangeably.

Chapter 5. Trigonometry

– The arcsine function is developed by limiting the sine function to $[-\pi/2, \pi/2]$

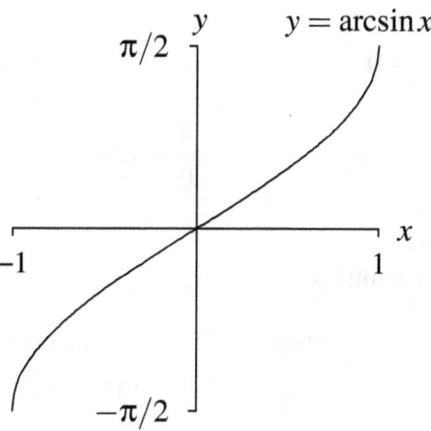

– The arccosine function is developed by limiting the cosine function to $[0, \pi]$

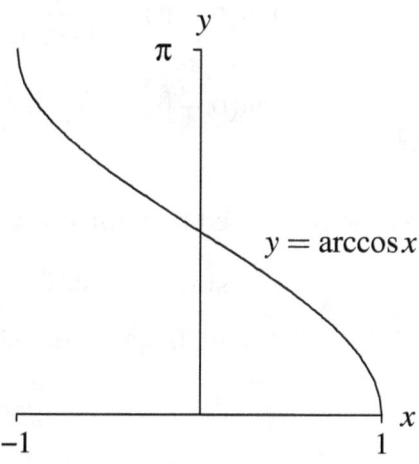

– The arctangent function is developed by limiting the tangent function to $(-\pi/2, \pi/2)$

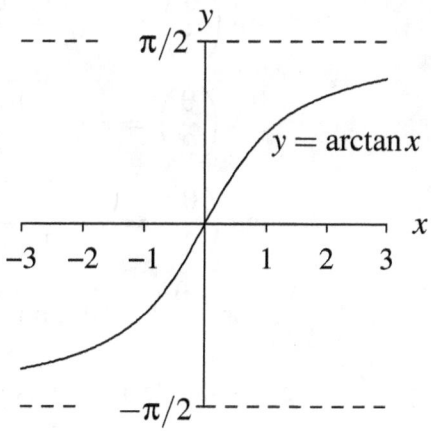

- Equations involving trigonometric functions can be solved using standard algebraic operations, for example,

$$2\sin x - 1 = 0 \quad \Rightarrow \quad 2\sin x = 1 \quad \Rightarrow \quad \sin x = \frac{1}{2}$$

$$\Rightarrow \quad x = \frac{\pi}{6} + (2\pi)n \quad \text{or} \quad x = \frac{5\pi}{6} + (2\pi)n$$

where n is an integer

- Sum and difference formulas

$$\sin(\alpha + \beta) = \sin\alpha\cos\beta + \cos\alpha\sin\beta$$
$$\sin(\alpha - \beta) = \sin\alpha\cos\beta - \cos\alpha\sin\beta$$
$$\cos(\alpha + \beta) = \cos\alpha\cos\beta - \sin\alpha\sin\beta$$
$$\cos(\alpha - \beta) = \cos\alpha\cos\beta + \sin\alpha\sin\beta$$
$$\tan(\alpha + \beta) = \frac{\tan\alpha + \tan\beta}{1 - \tan\alpha\tan\beta}$$
$$\tan(\alpha - \beta) = \frac{\tan\alpha - \tan\beta}{1 + \tan\alpha\tan\beta}$$

- Double-angle formulas can be derived by letting $\alpha = \beta = \theta$ in the sum formulas above

$$\sin 2\theta = 2\sin\theta\cos\theta$$
$$\cos 2\theta = \cos^2\theta - \sin^2\theta$$
$$\tan 2\theta = \frac{2\tan\theta}{1 - \tan^2\theta}$$

- Half-angle formulas

$$\sin\left(\frac{\theta}{2}\right) = \pm\sqrt{\frac{1 - \cos\theta}{2}}$$

$$\cos\left(\frac{\theta}{2}\right) = \pm\sqrt{\frac{1 + \cos\theta}{2}}$$

$$\tan\left(\frac{\theta}{2}\right) = \pm\sqrt{\frac{1 - \cos\theta}{1 + \cos\theta}}$$

The choice of sign for $\sin\left(\frac{\theta}{2}\right)$, $\cos\left(\frac{\theta}{2}\right)$, and $\tan\left(\frac{\theta}{2}\right)$ is determined by the quadrant associated with the angle $\frac{\theta}{2}$. The half-angle formulas can be derived from the double-angle formulas.

Chapter 5. Trigonometry

- The law of sines and the law of cosines apply to *oblique triangles*, which are triangles that do not contain a right angle

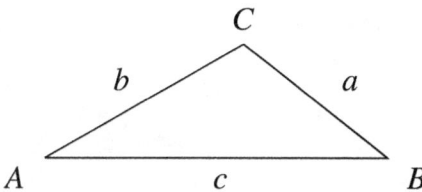

- *Law of sines:* for a triangle with sides of length a, b, and c, and corresponding opposite angles A, B, and C,

$$\frac{a}{\sin A} = \frac{b}{\sin B} = \frac{c}{\sin C}$$

or its reciprocal form

$$\frac{\sin A}{a} = \frac{\sin B}{b} = \frac{\sin C}{c}$$

- *Law of cosines:* for a triangle with sides of length a, b, and c, and angle C opposite the side of length c,

$$c^2 = a^2 + b^2 - 2ab\cos C$$

Analogous results exist for angles A and B. The law of cosines can be thought of as a generalization of the Pythagorean theorem for angles C that can be any angle measure (not just 90°).

- Polar coordinate system

  - The polar coordinate system is an alternative to the rectangular coordinate system
  - The polar coordinate system is centered about a point O known as the *pole* or *origin*
  - The ray directed eastward from the pole is known as the *polar axis*
  - Each point in the polar coordinate system can be determined by its *polar coordinates* (r, θ), where r is the directed distance from the pole to the point, and θ is the directed angle measured counterclockwise from the polar axis to the line segment connecting the pole to the point

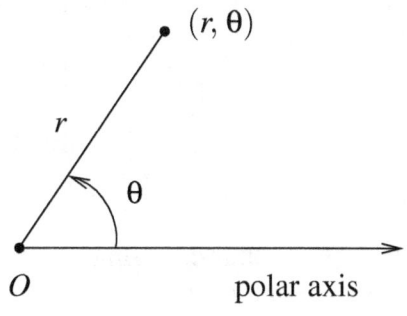

- Graph paper in the rectangular coordinate system consists of grid lines associated with constant values of x and y; graph paper in the polar coordinate system consists of circles and lines associated with constant values of r and θ

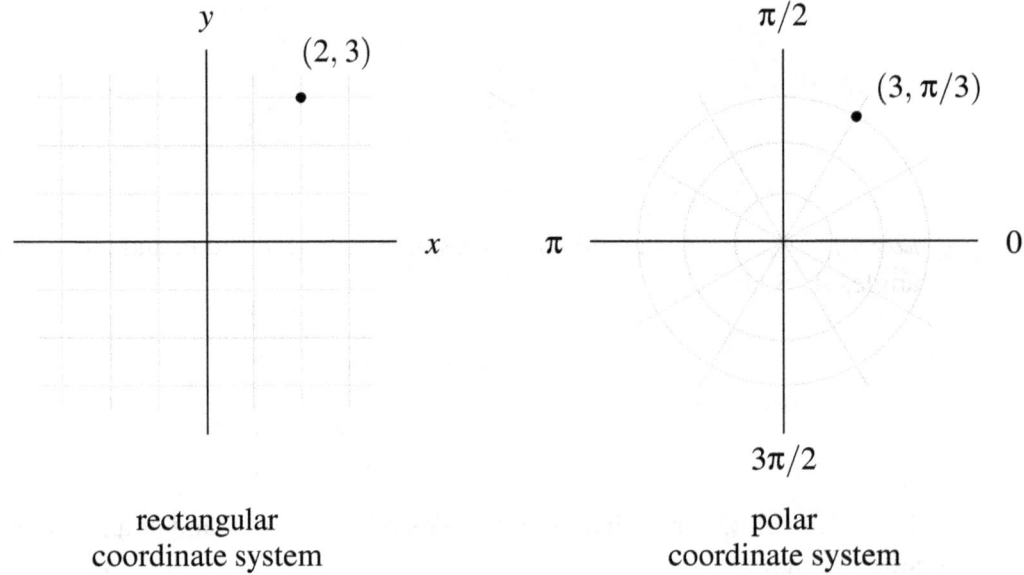

rectangular coordinate system

polar coordinate system

- A point (r, θ) in the polar coordinate system is typically plotted by first finding the direction of the directed angle θ, then plotting the point at this angle at the directed distance r. For example, the points

$$(2, \pi/3), (-3, \pi/4), (-2.6, -30°)$$

are plotted below on a grid consisting of concentric circles of radius 1, 2, and 3 and lines every $60°$.

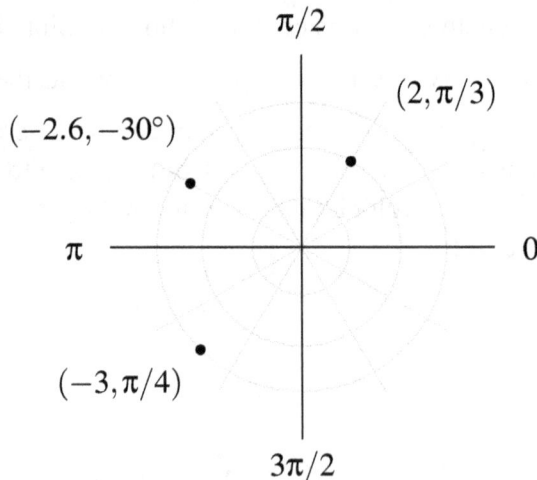

- A point (x, y) in the rectangular coordinate system has a unique representation; a point (r, θ) in the polar coordinate system does not have a unique representation. The four

Chapter 5. Trigonometry

ordered pairs in the polar coordinate system

$$\left(7, \frac{\pi}{4}\right), \left(7, \frac{9\pi}{4}\right), \left(-7, \frac{5\pi}{4}\right), \left(-7, -\frac{3\pi}{4}\right)$$

for example, all correspond to the same point.

– Converting between the systems

For a point (x, y) in the rectangular coordinate system that corresponds to a point (r, θ) in the polar coordinate system, the first three trigonometric functions are

$$\cos\theta = \frac{x}{r} \qquad \sin\theta = \frac{y}{r} \qquad \tan\theta = \frac{y}{x}$$

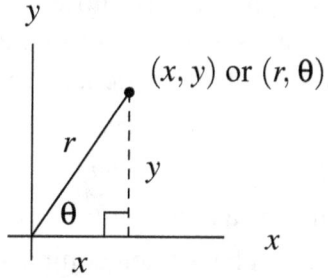

* Converting from the polar coordinate system to the rectangular coordinate system

$$x = r\cos\theta \qquad y = r\sin\theta$$

* Converting from the rectangular coordinate system to the polar coordinate system

$$r^2 = x^2 + y^2 \qquad \tan\theta = \frac{y}{x}$$

– Polar equations

* Polar equations typically express r as a function of θ
* Two of the simplest polar equations are assigning r and θ to constants. For example, $r = 3$ is a circle centered at the pole with radius 3, and $\theta = \pi/4$ is a line through the pole with inclination $\pi/4$, which corresponds to a slope of 1.
* Polar equations can be plotted by hand by making a table of values of θ and r. For example, $r = 2\cos\theta$ can be plotted with the table below.

θ	0	30°	60°	90°	120°	150°	180°	210°	240°	...
r	2	$\sqrt{3}$	1	0	-1	$-\sqrt{3}$	-2	$-\sqrt{3}$	-1	...

This is the equation of a circle in the polar coordinate system.

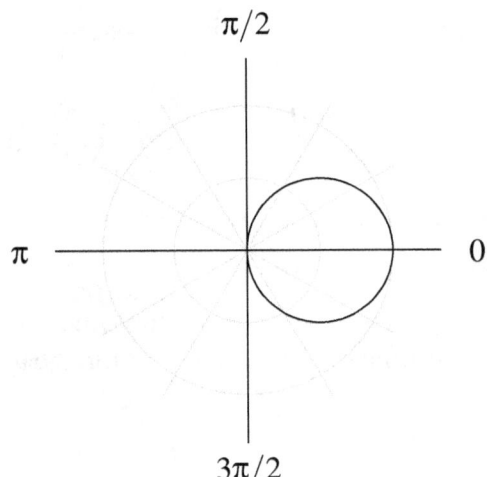

* Certain graphs (particularly those with a great deal of symmetry about the origin) are more easily expressed in polar form than rectangular form and are given colorful names by mathematicians, for example:
  - $r = 4\cos 3\theta$ is a *rose curve*
  - $r = 1 - 2\cos\theta$ is a *limaçon*
  - $r^2 = 9\sin 2\theta$ is a *lemniscate*

* Conic sections can be written compactly in the polar coordinate system by placing a focus at the pole and using the form

$$r = \frac{ep}{1 \pm ep\cos\theta} \qquad \text{or} \qquad r = \frac{ep}{1 \pm ep\sin\theta}$$

where $|p|$ is the distance between the focus and the directrix and $e > 0$ is the eccentricity ($0 < e < 1$ for an ellipse, $e = 1$ for a parabola, and $e > 1$ for a hyperbola). In these equations, the eccentricity e is not necessarily the constant $e = 2.718281828459045\ldots$.

Chapter 5. Trigonometry

Exercises

5.1 Convert the radian measure $\pi/9$ to degrees.

5.2 Convert the angle $108°$ to radian measure.

5.3 Find $\cos \pi$.

5.4 Find $\sin(-17\pi)$.

5.5 Find $\csc 45°$.

5.1 Convert the radian measure $\pi/9$ to degrees.

Multiplying the radian measure by $180°/\pi$ gives

$$\frac{\pi}{9} = \frac{\pi}{9} \cdot \frac{180°}{\pi} = 20°.$$

5.2 Convert the angle $108°$ to radian measure.

Multiplying the angle in degree measure by $\pi/180°$ gives

$$108° = 108° \cdot \frac{\pi}{180°} = \frac{3\pi}{5}.$$

5.3 Find $\cos \pi$.

The angle π, or $180°$, corresponds to the point $(-1, 0)$ on the winding function. Since the cosine function is the first component,

$$\cos \pi = -1.$$

5.4 Find $\sin(-17\pi)$.

Since -17π is eight and a half clockwise revolutions on the winding function, this angle corresponds to the point $(-1, 0)$. Since the sine function is the second component,

$$\sin(-17\pi) = 0.$$

5.5 Find $\csc 45°$.

The angle $45°$, or $\pi/4$, corresponds to the point $(1/\sqrt{2}, 1/\sqrt{2})$ on the winding function. So

$$\sin 45° = \frac{1}{\sqrt{2}},$$

which means that the cosecant of $45°$ is the reciprocal of the sine function at $45°$:

$$\csc 45° = \frac{1}{\sin 45°} = \sqrt{2}.$$

Chapter 5. Trigonometry

5.6 Find $\tan\left(\dfrac{2\pi}{3}\right)$.

5.7 Find $\sec\left(\dfrac{-5\pi}{4}\right)$.

5.8 If $\cos\theta < 0$ and $\sin\theta > 0$, which quadrant contains the terminal side of the angle θ?

5.9 For the triangle below, write x in terms of r and θ.

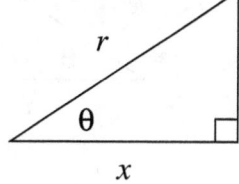

5.6 Find $\tan\left(\dfrac{2\pi}{3}\right)$.

The angle $2\pi/3$, or $120°$, corresponds to the point $(-1/2, \sqrt{3}/2)$ on the winding function. So

$$\sin\left(\frac{2\pi}{3}\right) = \frac{\sqrt{3}}{2} \qquad \text{and} \qquad \cos\left(\frac{2\pi}{3}\right) = -\frac{1}{2},$$

which means that the tangent of $2\pi/3$ is

$$\tan\left(\frac{2\pi}{3}\right) = \frac{\sin\left(\frac{2\pi}{3}\right)}{\cos\left(\frac{2\pi}{3}\right)} = \frac{\frac{\sqrt{3}}{2}}{-\frac{1}{2}} = -\sqrt{3}.$$

5.7 Find $\sec\left(\dfrac{-5\pi}{4}\right)$.

Since the secant function is the reciprocal of the cosine function,

$$\sec\left(\frac{-5\pi}{4}\right) = \frac{1}{\cos(-5\pi/4)} = \frac{1}{\cos(3\pi/4)} = \frac{1}{-1/\sqrt{2}} = -\sqrt{2}.$$

5.8 If $\cos\theta < 0$ and $\sin\theta > 0$, which quadrant contains the terminal side of the angle θ?

On the winding function, the cosine is the first component and the sine is the second component. Thus, when the first component is negative and the second component is positive, the angle θ has a terminal side that falls in the *second* quadrant.

5.9 For the triangle below, write x in terms of r and θ.

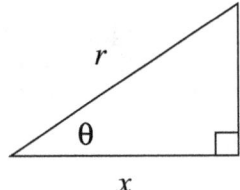

From the perspective of the angle θ, the length of the *adjacent* side x and the length of the *hypotenuse* r are known, so by the definition of the cosine function

$$\cos\theta = \frac{x}{r}.$$

Multiplying both sides of this equation by r gives

$$x = r\cos\theta.$$

Chapter 5. Trigonometry

5.10 Find the *amplitude* of the function $y = f(x) = 8 - 3\sin(2x - \pi)$.

5.11 Find the *range* of the function $y = f(x) = \sec x$.

5.12 If $\tan\theta = \sqrt{3}$, find the two possible values of $\sin\theta$.

5.13 Write $(\cos x)(\csc x)$ using a single trigonometric function.

5.14 Find $\cos(\arctan 8)$.

5.10 Find the *amplitude* of the function $y = f(x) = 8 - 3\sin(2x - \pi)$.

The amplitude is the absolute value of the coefficient of the sine function, which is 3.

5.11 Find the *range* of the function $y = f(x) = \sec x$.

The secant function is the reciprocal of the cosine function, that is,

$$y = f(x) = \sec x = \frac{1}{\cos x}.$$

Because the range of the cosine function is $[-1, 1]$, the reciprocal must be greater than or equal to 1 or less than or equal to -1. So the range of $y = f(x) = \sec x$ is

$$\{y \mid y \in \mathbb{R}, y \leq -1 \text{ or } y \geq 1\}.$$

5.12 If $\tan\theta = \sqrt{3}$, find the two possible values of $\sin\theta$.

The tangent is positive in the first and third quadrants. The angles θ that satisfy $\tan\theta = \sqrt{3}$ are $\pi/3$ and $4\pi/3$. So the sine of θ could be either

$$\sin\left(\frac{\pi}{3}\right) = \frac{\sqrt{3}}{2} \quad \text{or} \quad \sin\left(\frac{4\pi}{3}\right) = -\frac{\sqrt{3}}{2}.$$

5.13 Write $(\cos x)(\csc x)$ using a single trigonometric function.

Since the cosecant function is the reciprocal of the sine function,

$$(\cos x)(\csc x) = \cos x \cdot \frac{1}{\sin x} = \frac{\cos x}{\sin x} = \cot x.$$

5.14 Find $\cos(\arctan 8)$.

The tangent function is positive in the first and third quadrants. The arctangent function with a positive argument returns an angle in the first quadrant. Drawing a triangle in the first quadrant with an opposite side of length 8 and an adjacent side of length 1 will have a hypotenuse of length $\sqrt{8^2 + 1^2} = \sqrt{65}$ by the Pythagorean theorem. So the cosine of this angle is the adjacent side divided by the hypotenuse, which is

$$\cos(\arctan 8) = \frac{1}{\sqrt{65}}.$$

Chapter 5. Trigonometry

5.15 If $y = f(x) = \sin\left(x + \dfrac{\pi}{2}\right)$, find $f(\pi)$.

5.16 If $\cos\theta = \dfrac{1}{2}$ on $0 < \theta < \dfrac{\pi}{2}$, find θ.

5.17 Find $\cos\theta$ for the triangle shown below.

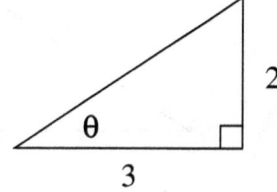

5.18 Find the *range* of the function $y = f(x) = 8 - 3\sin(2x - \pi)$.

5.19 Simplify $(\sin\theta)(\cos\theta)(\cot\theta)(\sec\theta)$.

5.15 If $y = f(x) = \sin\left(x + \dfrac{\pi}{2}\right)$, find $f(\pi)$.

The value of $f(\pi)$ is
$$f(\pi) = \sin\left(\pi + \dfrac{\pi}{2}\right) = \sin\left(\dfrac{3\pi}{2}\right).$$

Since $3\pi/2$ is three quarters of a counterclockwise revolution on the winding function, this angle corresponds to the point $(0, -1)$. Since the sine function is the second component,
$$f(\pi) = \sin\left(\dfrac{3\pi}{2}\right) = -1.$$

5.16 If $\cos\theta = \dfrac{1}{2}$ on $0 < \theta < \dfrac{\pi}{2}$, find θ.

Since the cosine of θ is the first component of any point on the winding function, the point $(1/2, \sqrt{3}/2)$ corresponds to $\pi/3$ or $60°$. So the angle θ is $\pi/3$ or $60°$.

5.17 Find $\cos\theta$ for the triangle shown below.

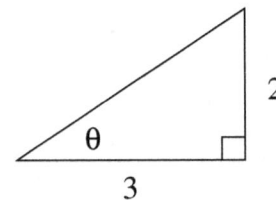

Using the Pythagorean Theorem, the length of the hypotenuse c can be found by solving
$$3^2 + 2^2 = c^2$$
for c yielding $c = \sqrt{13}$. Since the cosine is the ratio of the adjacent side to the hypotenuse,
$$\cos\theta = \dfrac{3}{\sqrt{13}}.$$

5.18 Find the *range* of the function $y = f(x) = 8 - 3\sin(2x - \pi)$.

The range is the set of all possible y values, which is $\{y \mid y \in \mathbb{R}, 5 \le y \le 11\}$.

5.19 Simplify $(\sin\theta)(\cos\theta)(\cot\theta)(\sec\theta)$.

By reducing all of the trigonometric functions to their definition in terms of sines and cosines:
$$(\sin\theta)(\cos\theta)(\cot\theta)(\sec\theta) = (\sin\theta)(\cos\theta)\left(\dfrac{\cos\theta}{\sin\theta}\right)\left(\dfrac{1}{\cos\theta}\right) = \cos\theta.$$

One fine point here is that the left-hand side of the equation is not defined at $\theta = 0, \pm\pi/2, \pm\pi, \ldots$, but the right-hand side of the equation is defined for all θ.

5.20 Find arccos($-1/2$).

5.21 Find arctan $\left(\sqrt{3}\right)$.

5.22 For the triangle shown below, write $(\tan\alpha)(\sin\beta)$ in terms of a, b, and c.

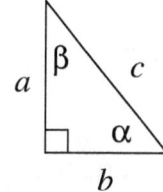

5.23 Write a function that describes the graph below (only one period is shown).

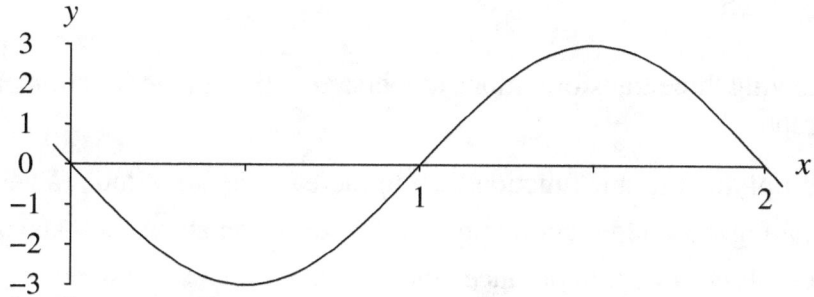

5.20 Find $\arccos(-1/2)$.

The angle on $[0, \pi]$ whose cosine is $-1/2$ is $2\pi/3$ or $120°$.

5.21 Find $\arctan(\sqrt{3})$.

The angle on $[-\pi/2, \pi/2]$ whose tangent is $\sqrt{3}$ is $\pi/3$ or $60°$.

5.22 For the triangle shown below, write $(\tan\alpha)(\sin\beta)$ in terms of a, b, and c.

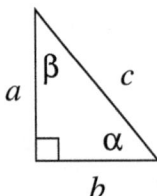

Using the definitions of the tangent and sine functions, the expression can be written as

$$(\tan\alpha)(\sin\beta) = \frac{a}{b} \cdot \frac{b}{c} = \frac{a}{c}.$$

5.23 Write a function that describes the graph below (only one period is shown).

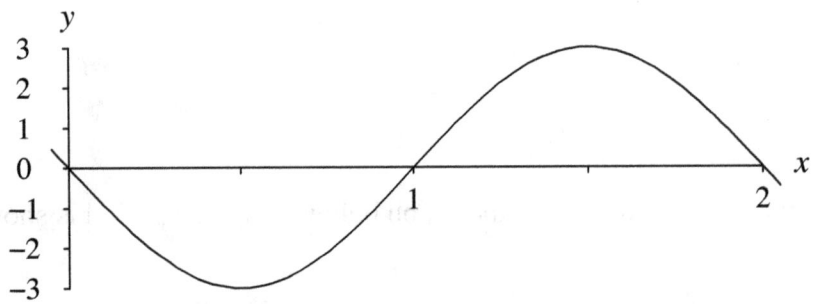

The following three transformations associated with the sine function can be used to arrive at the graph:

- multiplying the sine function by 3 to increase the amplitude (a vertical stretch),
- negating the sine function to produce a reflection about the x-axis,
- multiplying x by π to produce a horizontal shrink.

So one way to express the function is

$$y = -3\sin(\pi x).$$

An alternative solution using a shifted cosine function is

$$y = 3\cos\left(\pi x + \frac{\pi}{2}\right).$$

5.24 Find the exact value of $\cos 15°$.

5.25 If $3\tan^2 x + 2\sec^2 x = 9$, find $\tan x$.

5.24 Find the exact value of $\cos 15°$.

Solution 1. (Difference formula) Using the difference formula for the cosine function

$$\begin{aligned}\cos 15° &= \cos(60° - 45°) \\ &= \cos 60° \cos 45° + \sin 60° \sin 45° \\ &= \frac{1}{2} \cdot \frac{\sqrt{2}}{2} + \frac{\sqrt{3}}{2} \cdot \frac{\sqrt{2}}{2} \\ &= \frac{\sqrt{2} + \sqrt{6}}{4}.\end{aligned}$$

Solution 2. (Half-angle formula) Using the half-angle formula for the cosine function

$$\cos 15° = \cos\left(\frac{30°}{2}\right) = \pm\sqrt{\frac{1+\cos 30°}{2}} = \sqrt{\frac{1+\sqrt{3}/2}{2}} = \sqrt{\frac{1}{2} + \frac{\sqrt{3}}{4}}.$$

One can show that these two solutions are equal by comparing their squares.

5.25 If $3\tan^2 x + 2\sec^2 x = 9$, find $\tan x$.

Dividing the Pythagorean identity

$$\sin^2 x + \cos^2 x = 1$$

by $\cos^2 x$ gives the identity

$$\tan^2 x + 1 = \sec^2 x.$$

Using this identity to simplify the original equation yields

$$3\tan^2 x + 2\left(\tan^2 x + 1\right) = 9.$$

Collecting like terms,

$$5\tan^2 x + 2 = 9.$$

Subtracting two from each side of this equation gives

$$5\tan^2 x = 7.$$

Dividing both sides of this equation by 5,

$$\tan^2 x = \frac{7}{5}.$$

Taking the square root of both sides of this equation,

$$\tan x = \pm\sqrt{\frac{7}{5}}.$$

Chapter 5. Trigonometry

5.26 Solve

$$\frac{2\cos\left(\frac{2\pi}{3}\right) + \sin x}{\tan\left(\frac{3\pi}{4}\right)\cos\left(\frac{\pi}{7}\right) - \sin\left(\frac{\pi}{3}\right)} = 0$$

for x on the interval $0 \leq x \leq 2\pi$.

5.27 Solve

$$2\cos^2 x + 3\sin x = 3$$

for x on the interval $0 \leq x < 2\pi$.

5.26 Solve
$$\frac{2\cos\left(\frac{2\pi}{3}\right)+\sin x}{\tan\left(\frac{3\pi}{4}\right)\cos\left(\frac{\pi}{7}\right)-\sin\left(\frac{\pi}{3}\right)}=0$$
for x on the interval $0 \leq x \leq 2\pi$.

For this equation to be satisfied, the numerator must be 0. So solve
$$2\cos\left(\frac{2\pi}{3}\right)+\sin x = 0.$$

The angle $2\pi/3$ has an associated point $\left(-\frac{1}{2}, \frac{\sqrt{3}}{2}\right)$ on the winding function. So the cosine of $2\pi/3$ is $-1/2$. Thus, the equation reduces to
$$-1+\sin x = 0$$
or
$$\sin x = 1.$$

The angle on the interval $0 \leq x \leq 2\pi$ that solves this equation is $x = \pi/2$ or $x = 90°$.

5.27 Solve
$$2\cos^2 x + 3\sin x = 3$$
for x on the interval $0 \leq x < 2\pi$.

Using the Pythagorean identity $\sin^2 + \cos^2 = 1$, the equation can be written in terms of $\sin x$ as
$$2\left(1-\sin^2 x\right) + 3\sin x - 3 = 0.$$

Simplifying produces a quadratic function in $\sin x$:
$$2\sin^2 x - 3\sin x + 1 = 0,$$
which can be factored as
$$(2\sin x - 1)(\sin x - 1) = 0.$$

For this equation to hold, either
$$2\sin x - 1 = 0 \qquad \text{or} \qquad \sin x - 1 = 0$$
must hold. Equivalently,
$$\sin x = \frac{1}{2} \qquad \text{or} \qquad \sin x = 1.$$

The first equation is satisfied at $x = \pi/6$ and $x = 5\pi/6$. The second equation is satisfied at $x = \pi/2$.

Chapter 5. Trigonometry

5.28 Find all solutions of
$$5(\cos 2x)(2\cos x + 1) = 0$$
on $0 < x < \pi$.

5.29 For a triangle with side lengths $a = 2$, $b = 3$, and $c = 4$, find $\cos A$, where A is the angle opposite the side of length $a = 2$.

5.30 If $y = f(x) = \arcsin\left(x^2 + \dfrac{7}{16}\right)$, find $f\left(\dfrac{1}{4}\right)$.

5.28 Find all solutions of
$$5(\cos 2x)(2\cos x + 1) = 0$$
on $0 < x < \pi$.

Dividing both sides of this equation by 5 gives
$$(\cos 2x)(2\cos x + 1) = 0,$$
which is satisfied when
$$\cos 2x = 0 \quad \text{or} \quad 2\cos x + 1 = 0.$$

Examining the first equation, the cosine of an angle is 0 at $\pi/2$ and $3\pi/2$, so
$$2x = \frac{\pi}{2} \quad \text{or} \quad 2x = \frac{3\pi}{2}.$$

Solving for the angles x gives $x = \pi/4$ and $x = 3\pi/4$ on $0 < x < \pi$. The second equation can be arranged as
$$\cos x = -\frac{1}{2},$$
which is satisfied at $x = 2\pi/3$ on $0 < x < \pi$.

5.29 For a triangle with side lengths $a = 2$, $b = 3$, and $c = 4$, find $\cos A$, where A is the angle opposite the side of length $a = 2$.

Using the law of cosines,
$$a^2 = b^2 + c^2 - 2bc\cos A$$
or
$$2^2 = 3^2 + 4^2 - 2 \cdot 3 \cdot 4\cos A$$
or
$$4 = 9 + 16 - 24\cos A$$
or
$$-21 = -24\cos A.$$

Solving for $\cos A$ gives
$$\cos A = \frac{21}{24} = \frac{7}{8}.$$

5.30 If $y = f(x) = \arcsin\left(x^2 + \frac{7}{16}\right)$, find $f\left(\frac{1}{4}\right)$.

Evaluating $f(x)$ at $x = 1/4$,
$$f\left(\frac{1}{4}\right) = \arcsin\left(\frac{1}{16} + \frac{7}{16}\right) = \arcsin\left(\frac{8}{16}\right) = \arcsin\left(\frac{1}{2}\right) = \frac{\pi}{6}.$$

Chapter 5. Trigonometry

5.31 *Carefully* sketch
$$y = 2 + 2\cos\left(x - \frac{3\pi}{2}\right)$$
on $0 \leq x \leq 2\pi$.

5.32 When the sun is at an elevation of 30°, a building casts a 100 foot shadow. How tall is the building?

5.33 If $\tan\theta = \dfrac{a}{b}$ on $0 < \theta < \dfrac{\pi}{2}$ for $a > 0$ and $b > 0$, find $\sin\theta$.

5.31 *Carefully* sketch
$$y = 2 + 2\cos\left(x - \frac{3\pi}{2}\right)$$
on $0 \le x \le 2\pi$.

The phase shift of $3\pi/2$ shifts the cosine function $3\pi/2$ units to the right. (The period of the function is 2π.) The coefficient 2 before the cosine function is a vertical stretch that increases the amplitude to 2. Adding 2 results in an upward vertical shift. The graph of the function is shown below.

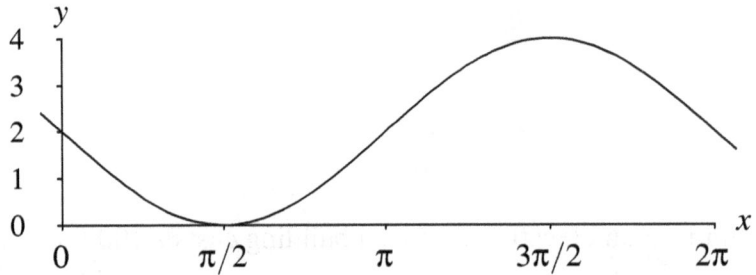

5.32 When the sun is at an elevation of $30°$, a building casts a 100 foot shadow. How tall is the building?

Let h be the height of the building. For the right triangle, created by the building and its shadow, the side opposite the $30°$ angle has length h and the side adjacent to the $30°$ angle has length 100. So by the definition of the tangent function,
$$\tan 30° = \frac{h}{100}.$$

Solving for h, the height of the building is
$$h = 100 \tan 30° = \frac{100}{\sqrt{3}} \cong 57.7 \text{ feet}.$$

5.33 If $\tan \theta = \dfrac{a}{b}$ on $0 < \theta < \dfrac{\pi}{2}$ for $a > 0$ and $b > 0$, find $\sin \theta$.

The triangle below satisfies $\tan \theta = a/b$ for the acute angle $0 < \theta < \pi/2$. The length of the hypotenuse is calculated by using the Pythagorean Theorem. Since the sine of θ is the ratio of the length of the opposite side to the length of the hypotenuse,
$$\sin \theta = \frac{a}{\sqrt{a^2 + b^2}}.$$

Chapter 5. Trigonometry

5.34 For the triangle below, find x.

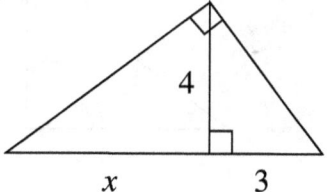

5.35 *Carefully* sketch $y = |\cos x|$ on $0 \leq x \leq 2\pi$.

5.34 For the triangle below, find x.

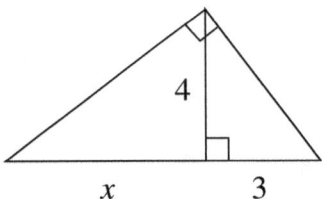

There are three right triangles in the figure. All three right triangles are similar because they contain common angles. To find the hypotenuse c of the smallest of these similar triangles, use the Pythagorean Theorem:
$$3^2 + 4^2 = c^2$$
or
$$25 = c^2.$$
Taking the positive square root of both sides of this equation, the length of the hypotenuse of the smallest triangle is $c = 5$. Since ratios of associated sides are equal for similar triangles, set the ratio of the length of the hypotenuse to the length of the shorter leg of the smallest triangle to the length of the hypotenuse to the length of the shorter leg of the largest triangle:
$$\frac{5}{3} = \frac{x+3}{5}.$$
Cross multiplying,
$$3x + 9 = 25.$$
Subtracting 9 from both sides of this equation gives
$$3x = 16.$$
Finally, dividing both sides of this equation by 3,
$$x = \frac{16}{3}.$$

5.35 *Carefully* sketch $y = |\cos x|$ on $0 \leq x \leq 2\pi$.

The absolute value reflects any point on the graph that is below the x-axis across the x-axis, resulting in the graph below.

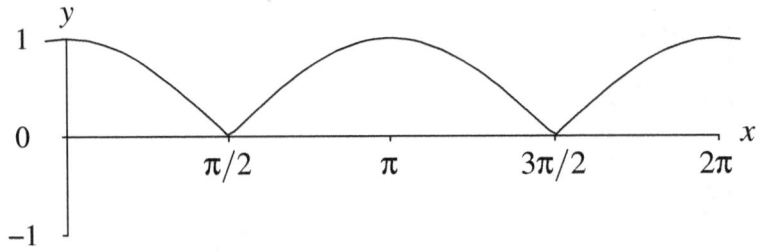

Chapter 5. Trigonometry

5.36 If $\cos\theta = -\dfrac{4}{5}$ and $\tan\theta > 0$, find $\csc\theta$.

5.37 Prove the identity
$$\frac{\sec\theta - 1}{1 - \cos\theta} = \sec\theta.$$

5.36 If $\cos\theta = -\dfrac{4}{5}$ and $\tan\theta > 0$, find $\csc\theta$.

The tangent of θ is positive in Quadrants I and III. Since the cosine of θ is negative, the angle θ must be in Quadrant III, and therefore the sine of θ and the cosecant of θ must be negative. Using the Pythagorean identity,

$$\sin^2\theta + \cos^2\theta = 1$$

or

$$\sin^2\theta + \left(-\frac{4}{5}\right)^2 = 1$$

or

$$\sin^2\theta + \frac{16}{25} = 1$$

or

$$\sin^2\theta = \frac{9}{25}.$$

Taking the square root of both sides of this equation,

$$\sin\theta = \pm\frac{3}{5}.$$

Discarding the positive value because θ is in the Quadrant III,

$$\sin\theta = -\frac{3}{5}.$$

Finally, the cosecant of θ is the reciprocal of the sine of θ:

$$\csc\theta = -\frac{5}{3}.$$

5.37 Prove the identity

$$\frac{\sec\theta - 1}{1 - \cos\theta} = \sec\theta.$$

Working with the left-hand side of the equation,

$$\begin{aligned}
\frac{\sec\theta - 1}{1 - \cos\theta} &= \frac{\frac{1}{\cos\theta} - 1}{1 - \cos\theta} \\
&= \frac{\frac{1-\cos\theta}{\cos\theta}}{1 - \cos\theta} \\
&= \frac{1}{\cos\theta} \\
&= \sec\theta.
\end{aligned}$$

5.38 For the triangle below, find $\cos 2\theta$.

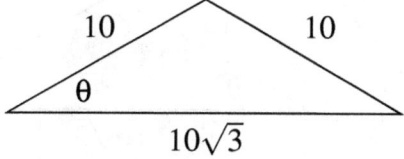

5.39 For the triangle below, find $\cos^2 \alpha + \cos^2 \beta$.

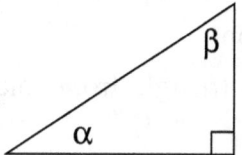

5.38 For the triangle below, find $\cos 2\theta$.

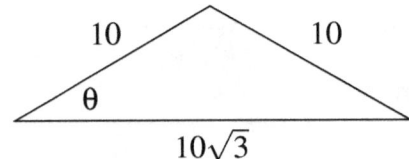

Solution 1. Since the two shorter sides of the triangle have equal lengths, their corresponding angles must be equal. Thus dropping a perpendicular bisector to the longer side creates two identical right triangles. Using the Pythagorean Theorem to determine the length of the vertical side b,
$$\left(5\sqrt{3}\right)^2 + b^2 = 10^2,$$
or
$$75 + b^2 = 100$$
which results in $b = 5$. Because the sine is the ratio of the length of the opposite side to the length of the hypotenuse,
$$\sin\theta = \frac{5}{10} = \frac{1}{2}.$$
So $\theta = 30°$. Finally,
$$\cos 2\theta = \cos 60° = \frac{1}{2}$$
by using the winding function.

Solution 2. Using the right triangle from solution 1 with $\theta = 30°$ and the double-angle formula,
$$\cos 2\theta = \cos^2\theta - \sin^2\theta = \left(\frac{\sqrt{3}}{2}\right)^2 - \left(\frac{1}{2}\right) = \frac{3}{4} - \frac{1}{4} = \frac{1}{2}.$$

5.39 For the triangle below, find $\cos^2\alpha + \cos^2\beta$.

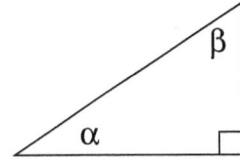

Since α and β sum to $\pi/2$ and cofunctions of complementary angles are equal,
$$\cos^2\alpha + \cos^2\beta = \cos^2\alpha + \cos^2\left(\frac{\pi}{2} - \alpha\right) = \cos^2\alpha + \sin^2\alpha = 1.$$

5.40 Find all angles x that satisfy
$$\cos\left(x - \frac{\pi}{5}\right) + 1 = 0.$$

5.41 *Carefully* plot a polar graph of $r = 1 - 2\cos\theta$.

5.40 Find all angles x that satisfy
$$\cos\left(x - \frac{\pi}{5}\right) + 1 = 0.$$
Subtracting 1 from both sides of the equation gives
$$\cos\left(x - \frac{\pi}{5}\right) = -1.$$
The only angle on the first rotation of the winding function that has a cosine of -1 is π. So
$$x - \frac{\pi}{5} = \pi.$$
Adding $\pi/5$ to both sides of this equation gives
$$x = \frac{6\pi}{5}.$$
Any angle that is a multiple of 2π more or less than this solution is also a solution, so the angles x that solve the equation are
$$x = \frac{6\pi}{5} + 2\pi n,$$
where n is any integer. Using degree measure, this can also be written as
$$x = 216° + 360°n,$$
where n is any integer.

5.41 *Carefully* plot a polar graph of $r = 1 - 2\cos\theta$.

The table below tabulates r for several values of θ.

θ	0	$\frac{\pi}{4}$	$\frac{\pi}{2}$	$\frac{3\pi}{4}$	π	$\frac{5\pi}{4}$	$\frac{3\pi}{2}$	$\frac{7\pi}{4}$	2π
r	-1	$1 - \sqrt{2}$	1	$1 + \sqrt{2}$	3	$1 + \sqrt{2}$	1	$1 - \sqrt{2}$	-1

The plot is the *limaçon* shown below.

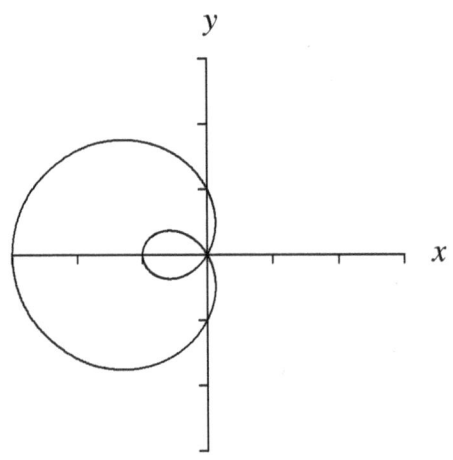

Chapter 6

Complex Numbers

Complex numbers extend the notion of the real number line from one dimension to two dimensions. These two dimensions account for the two parts of a number: the real part and the imaginary part. Complex numbers can be used to solve problems in applied mathematics that could not be solved with real numbers alone.

- Imaginary numbers
    - The *imaginary unit* is
    $$i = \sqrt{-1}$$
    - The first four powers of i are
    $$i^1 = i$$
    $$i^2 = -1$$
    $$i^3 = i^2 \cdot i = -i$$
    $$i^4 = i^2 \cdot i^2 = 1$$

    Higher powers of i cycle through these four values, for example, $i^{19} = -i$.
    - The general form of an *imaginary number* is
    $$bi$$
    where b is a nonzero real number
    - The square roots of negative numbers are imaginary numbers, for example,
    $$\sqrt{-36} = \sqrt{36} \cdot \sqrt{-1} = 6i$$
    or
    $$\sqrt{-2} \cdot \sqrt{-3} = \sqrt{2} \cdot \sqrt{-1} \cdot \sqrt{3} \cdot \sqrt{-1} = \sqrt{6}i^2 = -\sqrt{6}$$
    - If both a and b are negative, it is *not* true that $\sqrt{a} \cdot \sqrt{b} = \sqrt{ab}$, for example if $a = -2$ and $b = -3$, $\sqrt{a} \cdot \sqrt{b} = -\sqrt{6}$ but $\sqrt{ab} = \sqrt{6}$

- Complex numbers
    - The general form of a *complex number* is $a+bi$, where a and $b \neq 0$ are real numbers
        * when $a = 0$, the complex number reduces to the purely imaginary number bi
        * when $b = 0$, $a+bi$ reduces to the real number a
    - The set of complex numbers is denoted by C
    - Complex numbers can be thought of as keeping track of two numbers simultaneously: the real part a and the complex part b
    - Two complex numbers are equal if and only if their real and imaginary parts match
    - The *complex conjugate* of $a+bi$ is $a-bi$
    - Complex conjugates are used for simplifying quotients of complex numbers
    - Basic arithmetic operations
        * Addition
        $$(3-i) + (6+5i) = 9+4i$$
        * Subtraction
        $$(6+5i) - (3i) = 6+2i$$
        * Multiplication (via F.O.I.L.)
        $$(3-i) \cdot (6+5i) = 18 + 15i - 6i + 5 = 23 + 9i$$
        * Division (multiply numerator and denominator by the complex conjugate of the denominator in order to express the complex number in the form $a+bi$)
        $$\frac{3-i}{6+5i} = \frac{(3-i)(6-5i)}{(6+5i)(6-5i)} = \frac{18-15i-6i-5}{36-30i+30i+25} = \frac{13-21i}{61} = \frac{13}{61} - \frac{21}{61}i$$
    - A complex number multiplied by its complex conjugate yields
    $$(a+bi)(a-bi) = a^2 - abi + abi - b^2i^2 = a^2 + b^2$$
    which is a real number; in other words, it has no imaginary portion
    - A complex number $a+bi$ can be plotted in the *complex plane*, where the real portion a is plotted on the abscissa (the horizontal axis) and the imaginary portion b is plotted on the ordinate (the vertical axis)

- *Quadratic equations* of the form
$$ax^2 + bx + c = 0$$
have two roots (solutions) in the complex plane. These roots must be in one of the three classes given below.

Chapter 6. Complex Numbers

- The roots are real and equal, for example,

$$x^2 - 2x + 1 = 0$$
$$(x-1)^2 = 0$$
$$x = 1$$

which is a single root with multiplicity two

- The roots are real and unequal, for example,

$$x^2 - 3x + 2 = 0$$
$$(x-1)(x-2) = 0$$
$$x = 1 \quad \text{or} \quad x = 2$$

which are two distinct real roots

- The roots are complex conjugates, for example,

$$x^2 + 4 = 0$$
$$x^2 = -4$$
$$x = 2i \quad \text{or} \quad x = -2i$$

which are two distinct complex roots

- The graph of the quadratic function

$$y = f(x) = ax^2 + bx + c$$

is associated with the quadratic equation

$$ax^2 + bx + c = 0$$

which can be solved by the quadratic formula

$$x = \frac{-b \pm \sqrt{b^2 - 4ac}}{2a}$$

The *radicand* $b^2 - 4ac$ is known as the *discriminant*. The geometry associated with the three cases given above is described below.

- If the roots are real and equal, then the vertex of the parabola lies on the x-axis when $b^2 - 4ac = 0$

- If the roots are real and unequal, then the parabola intersects the x-axis twice when $b^2 - 4ac > 0$

- If the roots are complex conjugates, then the parabola never makes contact with the x-axis when $b^2 - 4ac < 0$

These three cases are illustrated in the three graphs below for quadratic functions with $a > 0$, which results in associated concave-up parabolas. Similar plots could be drawn for $a < 0$ and the associated concave-down parabolas.

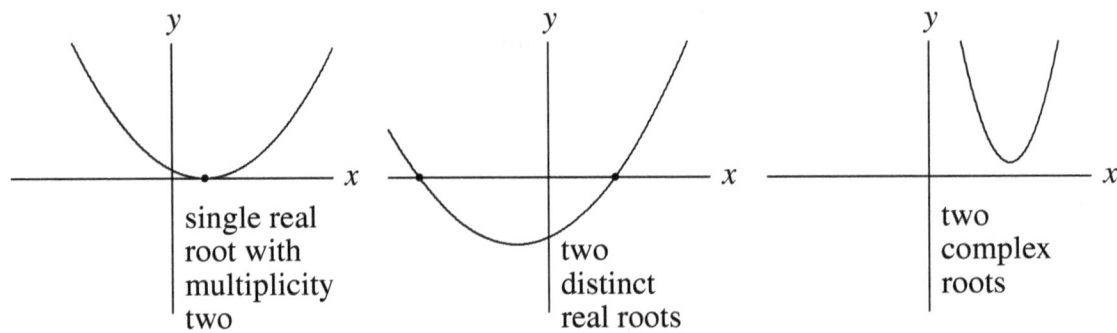

- Polynomial functions

    - Fundamental theorem of algebra: If $f(x)$ is a polynomial of degree $n > 0$, then it has at least one zero in the complex plane

    - The fundamental theorem of algebra implies that a polynomial of degree $n > 0$ has exactly n zeros in the complex plane when each root is counted up to its multiplicity (some of the roots might be real numbers)

    - Linear factorization theorem: If $f(x)$ is a polynomial of degree $n > 0$, then f has exactly n linear factors

    $$f(x) = a_n(x - c_1)(x - c_2)\ldots(x - c_n)$$

    where $c_1, c_2, \ldots, c_n$ are real or complex numbers

    - Complex zeros of a polynomial of degree $n > 1$ always occur in complex conjugate pairs; in other words, if $a + bi$ is a root of a polynomial, then so also is $a - bi$. For example,

    $$\begin{aligned} f(x) &= x^3 - x^2 + 9x - 9 \\ &= x^2(x - 1) + 9(x - 1) \\ &= (x^2 + 9)(x - 1) \\ &= (x + 3i)(x - 3i)(x - 1) \end{aligned}$$

    has zeros $-3i, 3i, 1$.

Chapter 6. Complex Numbers

Exercises

6.1 Write the *sum* of $3 - 2i$ and $4 + i$ in standard form.

6.2 Write the *product* of $3 - 2i$ and $4 + i$ in standard form.

6.3 Write the *quotient* of $3 - 2i$ and $4 + i$ in standard form.

6.4 Use the quadratic formula to find the zeros of $f(x) = x^2 - 2x + 9$.

6.5 Find the zeros of
$$f(x) = x^3 - x^2 + x - 1.$$

6.1 Write the *sum* of $3-2i$ and $4+i$ in standard form.

Adding the real and imaginary portions together, the sum is
$$(3-2i)+(4+i) = (3+4)+(-2i+i) = 7-i.$$

6.2 Write the *product* of $3-2i$ and $4+i$ in standard form.

Using F.O.I.L., the product is
$$(3-2i)\cdot(4+i) = 12+3i-8i-2i^2 = 14-5i.$$

6.3 Write the *quotient* of $3-2i$ and $4+i$ in standard form.

Multiplying the numerator and the denominator by the complex conjugate of the denominator gives
$$\frac{3-2i}{4+i} = \frac{3-2i}{4+i}\cdot\frac{4-i}{4-i} = \frac{(3-2i)(4-i)}{(4+i)(4-i)} = \frac{12-3i-8i+2i^2}{16-4i+4i-i^2} = \frac{10-11i}{17} = \frac{10}{17}-\frac{11}{17}i.$$

6.4 Use the quadratic formula to find the zeros of $f(x) = x^2 - 2x + 9$.

The quadratic formula
$$x = \frac{-b\pm\sqrt{b^2-4ac}}{2a}$$
with $a=1$, $b=-2$ and $c=9$ gives the complex conjugate pairs
$$x = \frac{2\pm\sqrt{4-(4)(1)(9)}}{2} = \frac{2\pm\sqrt{-32}}{2} = \frac{2\pm 4\sqrt{2}i}{2} = 1\pm 2\sqrt{2}i.$$

6.5 Find the zeros of
$$f(x) = x^3 - x^2 + x - 1.$$

Long division (or synthetic division) can be avoided by factoring x^2 out of the first two terms of the polynomial:
$$f(x) = x^3 - x^2 + x - 1 = x^2(x-1) + (x-1) = \left(x^2+1\right)(x-1) = (x+i)(x-i)(x-1).$$

Equating each of these factors to zero:
$$x+i=0 \qquad x-i=0 \qquad x-1=0.$$

So the polynomial has zeros $x=-i$, $x=i$, and $x=1$.

Chapter 6. Complex Numbers

6.6 Find all solutions to the equation
$$x^4 - x^2 - 56 = 0.$$

6.7 Consider the polynomial $y = f(x)$ whose graph is given below.

(a) What is the constant term of the polynomial?

(b) What is the smallest possible degree of this polynomial?

(c) What is the smallest possible number of complex zeros of this polynomial?

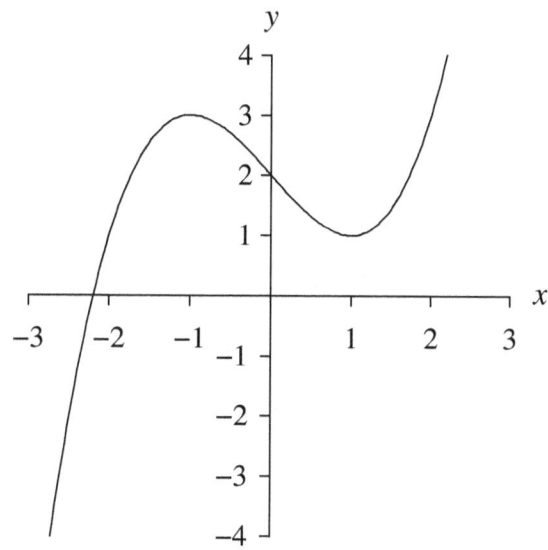

6.6 Find all solutions to the equation
$$x^4 - x^2 - 56 = 0.$$
The equation is quadratic in x^2, so it can be factored as
$$(x^2 - 8)(x^2 + 7) = 0$$
Treating each of the factors as the difference of squares, the equation can be further factored as
$$\left(x + \sqrt{8}\right)\left(x - \sqrt{8}\right)\left(x + \sqrt{-7}\right)\left(x - \sqrt{-7}\right) = 0$$
or
$$\left(x + 2\sqrt{2}\right)\left(x - 2\sqrt{2}\right)\left(x + \sqrt{7}i\right)\left(x - \sqrt{7}i\right) = 0.$$
So the four solutions to the equation are $x = -2\sqrt{2}$, $x = 2\sqrt{2}$, $x = -\sqrt{7}i$, and $x = \sqrt{7}i$.

6.7 Consider the polynomial $y = f(x)$ whose graph is given below.

(a) What is the constant term of the polynomial?

(b) What is the smallest possible degree of this polynomial?

(c) What is the smallest possible number of complex zeros of this polynomial?

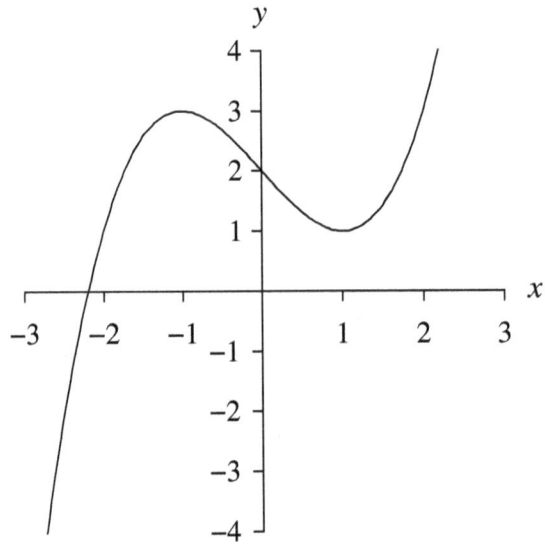

(a) Since the y-intercept of the graph is $y = 2$, the constant term in the polynomial is $a_0 = 2$.

(b) This polynomial could have three real roots (for a smaller value of the constant term), so it could be as low as a third-degree (cubic) polynomial.

(c) There is at least one real zero (between -3 and -2), so there are as few as two complex zeros.

Index

A

absolute value, *9*, 17–18, 34, 40–41
absolute value function, *87–88*, 121–122
acute angles, 65
acute triangles, 69
addition, 3
    properties, 27
adjacent angles, 65
algebra, 27–62
    equations, 32–35
    expanding expressions, 31–32
    exponents, 27–28
    expressions, 30–31
    factoring expressions, 32
    inequalities, 40–43
    logarithms, *29–30*, 45–48, 55–56, 59–60
    polynomials, *35–39*, 61–62
    properties of addition, 27
    properties of multiplication, 27
    radicals, 28–29
    simplifying expressions, 31
    word problems, *43–44*, 51–52
amplitude, *140*, 151–152, 155–156, 163–164
analytic geometry, 81–132
    Cartesian plane, 81
    classifying functions, 89
    combining functions, 90
    conic sections, *100–107*, 146
    direct and indirect variation, 107
    distance formula, 82–83
    exponential functions, 98–99
    functions, *85–88*, 109–122, 125–132
    graphs, 83–84
    intercepts, 84–85
    inverse functions, *90–91*, 111–112, 131–132
    linear functions (lines), *87–88*, *91–94*, 109–110
    logarithmic functions, *87–88*, *99–100*
    midpoint formula, 82
    one-to-one functions, *90*, 91, 98, 100, 129–130
    parametric equations, 107–108
    points, 82
    polynomial functions, *96–98*, 113–120, 125–128
    quadratic functions, 87–88, *95–96*, *101–103*, 115–116, 125–128
    rational functions, *98*, 119–120, 125–126
    symmetry, 83–84
    transforming functions, *89*, 115–116
angles, 64–66, 133–134
apothems, 68
arccosecant function (arccsc), 140
arccosine function (arccos)
    domain and range, 140
    graph, 141
arccotangent function (arccot), 140
arclength, 134
arcsecant function (arcsec), 140
arcsine function (arcsin)
    domain and range, 140
    graph, 141
arctangent function (arctan)
    domain and range, 140
    graph, 141
area, *73–75*, 79–80
area of regular polygons, 68
arithmetic, 1–26
    absolute value, *9*, 17–18
    arithmetic sequences and series, *12–13*, 19–20
    basic arithmetic operations, 3–4
    ceiling function, 9
    counting techniques, *14*, 21–26
    decimals, 7–8
    digits, 1, *14*
    exponents, *10–11*, 19–20
    factorial, *5*, 25–26

factors, 4
floor function, 9
fractions, *5–7*, 17–18
geometric sequences and series, 13–14
integers, 3
irrational numbers, 7
measurement systems, 11
modulo operator, 5
multiples, *4*, 21–22
number lines, 1–2
number systems, 15–16
order of operations, *11*, 17–18
percentages, *9–10*, 17–18, 21–22
prime numbers, 5
radicals, 11
rational numbers, 7
real numbers, 7
rounding, 8
scientific notation, 12
sets, 2–3
summation notation, 12
whole numbers, 1
arithmetic sequences and series, *12–13*, 19–20
associative property, 27
asymptotes, *89*, 98, 106, 119–122, 138–140
axis of symmetry, 95–96, 101–102

B
base, 10, 29, 99
basic arithmetic operations, 3–4
binary number system, 15
binomials, 31

C
cardinality, 2
Cartesian plane, 81
ceiling function, 9, 87–88
central angles of a circle, 72, 134
chord, 71
circles, *71–73*, 75, 79–80, *100–101*, 115–116, 145
circumference, 73
circumscribed circles, 72
classifying functions, 89
classifying polygons, 67
closed intervals, 40
coefficients, 31

cofunctions, 135
collinear points, 64
combinations, *14*, 23–26
combining functions, 90
common functions, 87–88
common logarithms, 29, 100
commutative property, 27
complement of a set, 3
complementary angle identities, *137*, 169–170
complementary angles, 66
completing the square, *39*, 49–50, 127–128
complex conjugates, 174–180
complex numbers, 173–180
    definition, 174
    graphs of quadratic functions, 175–176
    imaginary numbers, 173
    polynomial functions, 176
    quadratic equations, 174–175
composite numbers, 5
composition of functions, *90*, 131–132
compound interest, 17–18, *44*, 99, 119–120
concave polygons, 68
concave-down parabolas, *96*, 127–128, 176
concave-up parabolas, *95*, 115–116, 176
concentric circles, 73
cones
    surface area, 76
    volume, 76
congruent angles, 65
conic sections, *100–107*, 146
    circles, 100–101
    ellipses, *103–105*, 123–124
    hyperbolas, *105–107*, 121–122
    parabolas, *101–103*, 127–128
constant, 30
constant functions, 87–88
continuous functions, 86
converting from degrees to radians, *133*, 147–148
converting from radians to degrees, *133*, 147–148
convex polygons, 68
cosecant function (csc)
    definition, 135
    domain and range, 137
    graph, 140
cosine function (cos)

Index

definition, 135
domain and range, 137
graph, *138*, 155–156, 165–166
cotangent function (cot)
definition, 135
domain and range, 137
graph, 139
coterminal angles, 134
counting techniques, *14*, 21–26
cube root, 11
cube root function, 87–88
cubic functions, *87–88*, 113–114
cylinders
surface area, 75
volume, 76

D

decimal number system, 15
decimals, 7–8
decreasing functions, *87*, 98
degree measure for an angle, 64, 133–134
dependent variable, 85
Descartes' rule of signs, 98
diameter, 71–72
difference quotient, 94
digits, 1, *14*
direct and indirect variation, 107
directrix, *101–102*, 146
discriminant, 175
distance and travel problems, 44
distance formula, *82–83*, 100, 127–128
distributive property, 27
division, 4
domain, 30, 36, *85*, 86, 90–91, 109–110, 117–118, 129–132
double-angle formulas, *142*, 169–170

E

eccentricity, 146
of a hyperbola, 106
of an ellipse, 104–105
ellipses, *103–105*, 123–124, 146
endpoints, 63
equals sign (=), 2
equations, 32–35
equilateral triangles, *69*, 79–80

Euler's number, e, 30, *99*
even functions, *89*, 97, 111–112, 137
even/odd identities, 137
expanding expressions, 31–32
exponential functions, 87–88, *98–99*
exponents, *10–11*, 19–20, 27–28
expressions, 30–31
exterior angles of a polygon, 67
extraneous roots, 34

F

F.O.I.L., *31*, 35, 36, 174
factorial, *5*, 25–26
factoring expressions, 32
factoring quadratic equations, *39*, 49–50
factors, 4
floor function, 9, 87–88
focus, 101–107
45°, 45°, 90° triangles, 70, 134
fractions, *5–7*, 17–18
addition, 6
division, 6
lowest terms, 6
multiplication, 6
subtraction, 6
functions, *85–100*, 109–122, 125–132
arithmetic operations, 90
common functions, 87–88
composition, *90*, 131–132
definition, 85
inverse functions, *90–91*, 111–112, 131–132
notation, 85
one-to-one functions, *90*, 91, 98, 100, 129–130
zeros, *87*, 109–110, 113–114
fundamental theorem of algebra, 38, 176

G

general formula for a line, 92
geometric sequences and series, 13–14
geometry, 63–80
angles, 64–66
area, *73–75*, 79–80
circles, *71–73*, 75, 79–80
circumference, 73
line segments, *63–64*, 77–78
lines, 63–64

perimeter, *73*, 77–78
planes, 64
points, 63
polygons, 66–71
rays, 63–64
solid geometry, 75–76
grads, 134
graphing techniques, *83*, 107, 145–146
graphs, 83–84
graphs of quadratic functions, 87–88, *95–96*, *101–103*, 175–176
graphs of trigonometric functions, *137–140*, 155–156, 163–166
greater than (>), 2, 40
greater than or equal to (≥), 40
greatest common factor (GCF), 4
greatest integer function, 9, 87–88

H
half-angle formulas, *142*, 157–158
Heron's formula, 73
hexagons, 68
horizontal line test, *90*, 91, 129–130
hyperbolas, *105–107*, 121–122, 146
hypotenuse, *69*, 109–110, 134–135

I
identity function, *87–88*, 91
imaginary numbers, 173
imaginary unit, 173
improper fractions, 5
increasing functions, *87*, 98, 100
independent variable, 85
inequalities, *40–43*, 61–62
    involving linear expressions, 41
    involving polynomial expressions, 42
    involving quadratic expressions, *41*, 61–62
    involving rational expressions, 42
inherently quadratic equations, *39*, 61–62, 179–180
initial side of an angle, 133
inscribed angles of a circle, 72
integers, 3
intercepts, *84–85*, 87, 92, 96, 97, 123–124
interest problems, 17–18, *44*, 99, 119–120
interior angles of a polygon, 67
intermediate value theorem, 86

intersection of sets, 3
interval notation, 40
inverse functions, *90–91*, 111–112, 131–132
inverse trigonometric functions, *140–142*, 161–162
irrational numbers, 7
isosceles triangles, 69

L
law of cosines, *143*, 161–162
law of sines, 143
least common denominator (LCD), *6–7*, 17–18
least common multiple (LCM), *4*, 7, 17–18, 21–22
least integer function, 9, 87–88
lemniscates, 146
less than (<), 2, 40
less than or equal to (≤), 40
limaçons, *146*, 171–172
line segments, *63–64*, 77–78
linear equations, 38
linear factorization theorem, 176
linear functions (lines), 87–88, *91–94*, 109–110
lines, 63–64, 91–94
logarithmic functions, 87–88, *99–100*
logarithms, *29–30*, 45–48, 55–56, 59–60

M
major arcs of a circle, 72
major axis, 103–105
measure of an angle, 64
measurement systems, 11
midpoint formula, 82
minor arcs of a circle, 72
minor axis, 103–105
mixed numbers, 5
mixture problems, 44
modulo operator, 5
monomials, 31
multiples, *4*, 21–22
multiplication, 4
    properties, 27
multiplication rule, *14*, 21–22
multiplicity of a zero, *97*, 113–114

N
natural logarithms, 29, 100
negative numbers, 1

Index

nondecreasing functions, 87
nonincreasing functions, 87
not equals sign ($\neq$), 2
number lines, *1–2*, 81
number systems, 15–16

O

oblique triangles, 143
obtuse angles, 65
octagons, 68
octal number system, 15
odd functions, *89*, 97, 137
one-to-one functions, *90*, 91, 98, 100, 129–130
open intervals, 40
opposites, 2
order of operations, *11*, 17–18
orientation of a plane curve, 107
origin, 1, 81, 82, 143

P

parabolas, *95–96, 101–103*, 115–116, 127–128, 146, 175–176
parallel lines, 64, 93
parallelograms, 71, 74
parameters of trigonometric functions, 140
parametric equations, 107–108
PEMDAS, 11
pentagons, 67
percentages, *9–10*, 17–18, 21–22
perfect cubes, *11*, 23–24
perfect squares, *10*, 23–24
perimeter, *73*, 77–78
period of trigonometric functions, 137–140
permutations, 14
perpendicular lines, 64, 93
phase shift, *140*, 163–164
piecewise functions, *86*, 113–114, 129–132
place value, 1
plane curve, 107
planes, 64
plus or minus sign ($\pm$), 2
point–slope formula for a line, 92
points, 63, 82, 144–145
polar axis, 143
polar coordinate system, *143–146*, 171–172
polar equations, *145–146*, 171–172

pole, 143
polygons, 66–71
polynomial functions, *96–98*, 113–120, 125–128, *176*, 179–180
    constant term, 97, 179–180
    leading coefficient, 97
    limiting behavior, 97
    multiplicity of a zero, *97*, 113–114
    possible rational zeros, 98
    standard form, 96
polynomials, 35–39, 61–62
    basic operations, 35
    classification, 35–36
    constant term, 35, 179–180
    factoring, 36
    inherently quadratic equations, *39*, 61–62, 179–180
    leading coefficient, 35
    long division of, 37, 57–58, 97
    polynomial equations, 37
    quadratic equations, 38–39
    rational expressions, 36–37
    standard form, 35
    synthetic division of, *37–38*, 57–58, 97, 117–118
positive numbers, 1
power functions, 87–88
powers of trigonometric functions, 136
prime factorization, *5*, 17–18, 25–26
prime numbers, 5
principal root, 28
properties of addition, 27
properties of multiplication, 27
Pythagorean identities, *137*, 157–160, 167–168
Pythagorean theorem, *70*, 83, 143, 151–154, 163–166, 169–170

Q

quadrants, *81*, 117–118, 149–150
quadratic equations, *38–39*, 49–50, 61–62, 174–175
quadratic formula, *39*, 49–50, 177–178
quadratic functions, 87–88, *95–96, 101–103*, 115–116, 125–128, 175–176
    applications, 96, 103
    general form, 95

standard form, 95
quadrilaterals, 67, *70–71*
quotient identities, 137

R

radian measure for an angle, 133–134
radicals, 11, 28–29
radius, *71*, 79–80
range, *85*, *86*, 90–91, 117–118, 125–126, 129–130, 151–154
rational expressions, 36–37
rational functions, *98*, 119–120, 125–126
rational numbers, 7
rays, 63–64
real numbers, 7
reciprocal function, *87–88*, 89, 91
reciprocal identities, 136–137
reciprocals, 7
rectangles, 71, 74, 75
rectangular coordinate system, *81*, 144
rectangular solids
    surface area, 75
    volume, 76
reference boxes for hyperbolas, 106
reflections, *89*, 131–132, 165–166
reflective property of a parabola, 103
regular polygons, 68
remainder, *4*, 5, 8, *37*
repeating decimals, *8*, 19–20
rhombus, 71
right angles, 65
right triangles, *69–70*, 134–136
rigid transformations, *89*, 131–132
rise over run, *91*, 92
Roman numerals, 15–16
roots, 32
rose curves, 146
rounding, 8

S

scalene triangles, 69
scientific notation, 12
secant function (sec)
    definition, 135
    domain and range, 137
    graph, 139

secant line, 94
semicircles, 72
sets, 2–3
    cardinality, 2
    complement of, 3
    intersection of, 3
    subset, 2
    union of, 3
shifts, *89*, 131–132
similar triangles, *69*, 165–166
simple interest, 44
simplifying expressions, 31
sine function (sin)
    definition, 135
    domain and range, 137
    graph, *138*, 155–156
slope, *91–92*, 109–110
slope–intercept formula for a line, 92
SOH CAH TOA, 135
solid geometry, 75–76
solutions, 32
solving equations, *32–34*, 45–60
solving trigonometric equations, *142*, 159–162, 171–172
spheres
    surface area, 76
    volume, 76–78
square root, 11
square root function, *87–88*, 115–116
squares, *71*, 74, 75, 115–116
standard position of an angle, 133
straight angles, 65
subscripts, 13
subset, 2
subtraction, 3–4
sum and difference formulas, *142*, 157–158
summation notation, 12
superscripts, 10
supplementary angles, 66
surface area, *75–76*, 111–112
symmetry, 83–84
    about the x-axis, 84
    about the y-axis, 83, 89
    about the origin, 84, 89
systems of equations, *35*, 59–60, 125–126

Index

T

tangent function (tan)
    definition, 135
    domain and range, 137
    graph, 138
tangent line, 99
terminal side of an angle, 133
terminating decimals, 8
30°, 60°, 90° triangles, 70, 134
transforming functions, *89*, 115–116
translations, 89
transversal lines, 66
transverse axis, 105–107
trapezoids, *71*, 74
triangles, *67–70*, 109–110
    area, 73
    classification, 69
    sum of degree measures, 68
trigonometric functions, *134–135*, 147–172
trigonometric identities, 136–137
trigonometry, 133–172
    angles, 133–134
    double-angle formulas, *142*, 169–170
    45°, 45°, 90° triangles, 134
    graphs of trigonometric functions, *137–140*, 163–166
    half-angle formulas, *142*, 157–158
    inverse trigonometric functions, *140–142*, 161–162
    law of cosines, *143*, 161–162
    law of sines, 143
    parameters of trigonometric functions, 140
    polar coordinate system, *143–146*, 171–172
    powers of trigonometric functions, 136
    solving trigonometric equations, *142*, 159–162, 171–172
    sum and difference formulas, *142*, 157–158
    30°, 60°, 90° triangles, 134
    trigonometric functions, *134–135*, 147–172
    trigonometric identities, 136–137
    winding function, *135–136*, 147–172
trinomials, 31

U

union of sets, 3

unit circle, 101, 135
unit multipliers, *11*, 133

V

variables, 30
Venn diagrams, 2, 21–22
vertex, 101–107
vertex of an angle, 64
vertical angles, 66
vertical line test, *85*, 91
vertical lines, 93
volume, 76

W

whole numbers, 1
winding function, *135–136*, 147–172
word problems, *43–44*, 51–52
    classification, 44
    steps for solving, 43
work problems, *44*, 51–52

X

x-axis, 81
x-intercepts, *85*, 87, 96, 97, 123–124

Y

y-axis, 81
y-intercepts, *84*, 92, 96, 97, 123–124

Z

zeros of a function, *87*, 109–110, 113–114, 176–180